The Access Manual

The Authors

Ann Sawyer BA Dip. Arch. is an architect and independent access consultant. She works with architects, developers, building owners and facilities managers providing access consultancy services on new and existing buildings and environments. She advises on access and inclusive design, provides access appraisals and audits, advises on improvements to meet new legislative requirements and provides training in accessible design, auditing and the management of accessible environments. She has been involved in many prestigious new build and refurbishment projects for public and private sector clients and has advised on access to a wide variety of historic buildings.
Ann can be contacted on a.sawyer@blueyonder.co.uk

Keith Bright is Professor of Inclusive Environments at The University of Reading and Director of Keith Bright Consultancy, a company offering independent access advice and training. He was a serving member on BSI Committee B/209/8 involved in the preparation of BS8300: 2001. Keith is a Registered Access Consultant (NRAC), and a member of the Management Committee for the National Register of Access Consultants. He is also a member of the Built Environment Working Group (BEWG) set up under the Disabled Persons Transport Advisory Committee (DPTAC). He has undertaken a wide portfolio of research in areas relating to the design and management of inclusive accessible environments and is a regular speaker at seminars and conferences both nationally and internationally.
Keith can be contacted on k.t.bright@reading.ac.uk

The Access Manual

Auditing and managing inclusive built environments

Ann Sawyer
Architect and Access Consultant

and

Keith Bright
Registered Access Consultant and Professor of Inclusive Environments
The University of Reading

Blackwell
Publishing

© 2004 by Blackwell Publishing Ltd

Editorial Offices:
Blackwell Publishing Ltd, 9600 Garsington Road, Oxford OX4 2DQ, UK
    Tel: +44 (0)1865 776868
Blackwell Publishing Professional, 2121 State Avenue, Ames,
Iowa 50014-8300, USA
    Tel: +1 515 292 0140
Blackwell Publishing Asia, 550 Swanston Street, Carlton, Victoria 3053,
Australia
    Tel: +61 (0)3 8359 1011

First published 2004 by Blackwell Publishing Ltd
Reprinted 2004

Library of Congress Cataloging-in-Publication Data is available

ISBN 1-4051-0765-0

A catalogue record for this title is available from the British Library

Set in 12/15pt Optima
by Graphicraft Limited, Hong Kong
Printed and bound in Great Britain
by MPG Books, Bodmin, Cornwall

The publisher's policy is to use permanent paper from mills that operate a
sustainable forestry policy, and which has been manufactured from pulp
processed using acid-free and elementary chlorine-free practices. Furthermore,
the publisher ensures that the text paper and cover board used have met
acceptable environmental accreditation standards.

For further information on Blackwell Publishing, visit our website:
www.blackwellpublishing.com

Contents

'Good design should be for everyone'

Daniel Libeskind

Foreword

It is impossible to accurately describe how design takes place. It is part personal and part a question of balancing a broad and complex group of internal and external influences each with their own personal input. What I can say is that the success of any project is determined by the relationship between client, architect and the broader consultant team.

At RRP we engage with specialists from inception to completion. They contribute to and are instrumental in developing ideas into reality as we tread a balance between the inspirational and pragmatic. At the root of our architecture is a desire to be socially responsible, to innovate and to question. We aim to create open and expressive buildings that have clarity of organisation and a more sustainable agenda – exciting people places.

Access is embodied into our approach at all levels. Our scheme to reinvigorate the South Bank cultural centre in London in the mid-1990s revolved around the opening up of access to a complex spectacular for its contortions and failures in this respect. This was a clear case where poor access was synonymous with a poor image and a dearth of good people spaces.

Our design concept for the National Assembly for Wales was conceived around clear notions of openness and clarity, a clear relationship between the electorate and elected, a sustainable approach to a 21st century seat of democracy. Exemplar access is the key to achieving these goals. The biggest single driver to the development of the design concept has been the response to public consultation and specialist advice on access issues. Healthy criticism and debate of access issues is essential to turn a scheme from an idea into reality, and to achieve buildings accessible to all.

Ivan Harbour
Director, **Richard Rogers Partnership**

Acknowledgements

I would like to thank Vin Goodwin, for his advice, opinions and encouragement, the Centre for Accessible Environments for providing me with the opportunity to explore the issues that form the basis of this book, and Keith Bright, whose knowledge and enthusiasm have been invaluable. Thanks also to my family for giving support and inspiration.

Ann Sawyer
September 2003

I would like to express my thanks to all the members of the Research Group for Inclusive Environments at The University of Reading who, over a period of ten years, have contributed to the complex research activities on which many of the comments made in this book are based.

My special thanks go to Dr Geoffrey Cook for his valuable guidance on issues related to lighting, colour and acoustics, and for being an excellent working companion and good friend over the last ten years.

Special thanks also go to Ann Sawyer who has worked very hard to bring the dream of this publication to a reality.

Keith Bright
September 2003

Introduction

Buildings are designed for people to use – to give shelter, to house, for work and for play. An environment that is designed to be accessible, or inclusive, allows those activities to take place without restricting access to people with certain abilities only. Inclusive design does not disable the users; it enables independent and equal use.

For many owners, designers and managers of buildings and environments, meeting the needs of all users, and especially those with disabilities, can seem difficult. However, it is possible to address the needs of the great majority of users with design and management solutions that neither conflict with each other, nor are expensive or difficult to carry out. An inclusive approach requires designers and building managers to consider abilities rather than disabilities, and integrate a range of needs into one solution that can be used by everyone. The improvement in accessibility that can result from this approach will benefit all users of the built environment, not just disabled people.

Inclusive design

Design guidance is often based on the needs of a notional 'average' person; however, everyone varies from the average in some way. People differ in height, strength and dexterity. People have different visual and hearing abilities or may have respiratory impairments or reduced stamina. Some of these people will consider themselves 'disabled', some will not. Some will see their abilities as a natural result of ageing or maybe a temporary illness. Older people may have limited mobility; some may use wheelchairs, sticks or crutches. These and other aids such as spectacles and hearing aids will allow people to alter their abilities. Mobility may be affected by having to carry a child or heavy shopping bags or push a buggy. These are all usual, everyday capacities that should be catered for when designing buildings and environments. An inclusive approach accepts that people have a range of needs and leads to designs that allow the majority of people to use the built environment independently, safely and comfortably.

An example of this would be good, clear, effective signage, efficiently and sensibly used. This is not just needed by people with visual impairments; it is important for everyone and critical for those who are deaf or hard of hearing, people with learning difficulties or disabilities, older people or people whose first language is not that of the information given on the sign. Also, level, firm, surfaces, which benefit wheelchair users, will help parents with pushchairs, people using walking aids or those carrying luggage.

An inclusive approach to design and management does not deny that there are specific areas where particular assistance can be provided. Hearing enhancement systems, such as induction loops, or the provision of information in Braille are useful to certain building users. Specific provisions that meet particular needs are also part of inclusive design.

Principles of universal design

There are many different terms in use to describe inclusive design. Terms such as accessible, trans-generational, or universal all have similar and overlapping definitions; what they all have in common is that the needs of as many people as possible are considered, including parents with children, elderly people and people with disabilities. The Centre for Universal Design has produced a useful list of principles of universal design that can be applied to products and buildings (see Box I.1).

Costs and benefits

It is often thought that addressing the needs of everyone in new or existing buildings will lead to increased costs. However, careful consideration of the issues at design stage and good management throughout the life of a building can provide and maintain accessible environments at little, or no, extra cost. Designing buildings

1.1 Principles of universal design

1 Equitable – the design should be usable by people with diverse abilities and should appeal to all users.
2 Flexible – the design should cater for a wide range of individual preferences and abilities. This may mean some choice in methods of use (such as right or left handed access).
3 Simple and intuitive – use of the design should be easy to understand, regardless of experience, knowledge, language skills, or current concentration level.
4 Perceptible – the design communicates necessary information effectively to the user, regardless of ambient conditions or the user's sensory abilities.
5 Tolerance for error – the design minimises hazards and the adverse consequences of accidental or unintended actions.
6 Low physical effort – the design can be used efficiently and comfortably with a minimum of fatigue.
7 Size and space for approach and use – appropriate size and space is provided for approach, reach, manipulation and use, regardless of user's body size, posture or mobility.

and environments to be inclusive from the outset can also avoid the need for costly, and often unsightly, alterations later on.

Elements of the built environment, such as buildings, pedestrian areas and transport infrastructure, are with us for a long time, but their life is dynamic, not static, and there are often opportunities to improve accessibility. Linking improvements to maintenance or refurbishment programmes can help ensure that the work is done cost-effectively.

There are also financial benefits to be had from designing inclusively. The population is ageing; current estimates suggest that by 2015 nearly one in five of us will be over 65. Older people with higher disposable incomes are becoming a more important force in the market place, and with increased opportunities in employment, the spending power of people with disabilities will also grow. Service providers can increase and broaden their customer base by making their services, and the buildings that house them, accessible to everyone. Employers can benefit from the skills and abilities of people with disabilities by ensuring that their buildings and procedures are accessible.

Disability Discrimination Act 1995

With the introduction of the Disability Discrimination Act 1995 (DDA) the consideration of issues such as access and inclusive design has become even more important. The Act gives rights to people with disabilities with the intention of eliminating discrimination in the areas of:

- recruitment and employment;
- access to goods, facilities and services;
- buying or renting land or property.

The DDA applies to the whole of the UK and places duties on employers and service providers not to discriminate against people with disabilities. The DDA extends the definition of disability to cover a wide range of people including people with hearing and visual impairments, learning difficulties, mental illness and ambulant disabled people, such as those using walking aids or with arthritis. This definition may well be altered over time as cases come before the courts and as other legislation, such as the Human Rights Act, which expands the definition of disability even further, start to have an influence.

The provisions of the DDA are being introduced over a period of time. The employment provisions have been in force since 1996 and the provisions affecting service providers are due to be fully in place by 2004. From that date all service providers will have to make 'reasonable adjustments' to physical features of their premises to overcome physical barriers to access.

The DDA does not directly require buildings to be accessible to all disabled people and does not include standards for accessible building design; it is the services on offer within buildings that are the concern of the Act. However, it is critical that building owners and managers, facilities managers and those commissioning or designing new buildings or works to existing buildings consider the implications of the Act in relation to building design and use. This will involve anticipating the needs of all building users, some of whom will fit the definition of disabled under the Act, and designing and managing buildings accordingly. Knowledge of access audits and access management will be a crucial factor in determining how well this can be done and will allow an effective response to the new legal requirements.

As the DDA begins to have an effect, expectations of disabled people are likely to rise and what is currently acceptable in design terms may not be in the future. The letting and selling of property used by employers and service providers will also be affected, with inaccessible premises becoming less desirable as the implications of the Act are recognised.

British Standards and Building Regulations

The publication of BS 8300:2001 Design of buildings and their approaches to meet the needs of disabled people – Code of practice, and the subsequent revision of Part M of the Building Regulations, has influenced standards of good practice in inclusive design.

BS 8300:2001 gives detailed guidance on the design of domestic and non-domestic buildings. The guidance draws on research, commissioned by the Department of the Environment, Transport and the Regions in 1997 and 2001, into the access needs of people with disabilities. The research looked into ergonomic issues such as reach ranges and space requirements in order to assess the capabilities and needs of people in relation to the use of buildings.

The BS is the most comprehensive standard to date covering the needs of people with disabilities and makes recommendations that exceed those given in previous guidance. It is under regular review and may well change as more research is carried out. It is likely that it will be used as a benchmark when considering what is 'reasonable provision' in relation to the Disability Discrimination Act.

The new edition of Part M of the Building Regulations, due to come in to force in spring 2004, refers to new and existing buildings being accessible and usable by people, including parents with children, elderly people and people with disabilities. The dimensional criteria in the new edition of Part M have been updated and revised to take account of the guidance given in BS 8300:2001.

The standards and dimensions recommended in this book are generally in line with BS 8300:2001. The good practice guidance draws on other sources from this country and abroad, from user groups and from the wide experience of the authors.

Objectives of the manual

The manual covers the design, improvement, maintenance and management of accessible environments. The intention is to encourage designers, owners and managers of buildings to look at how they can provide and operate buildings, services and employment facilities in a way that allows independent and convenient use by everyone.

The manual is intended to enable people with responsibility for the design and management of the built environment to:

- be aware of the issues involved in accessibility;
- understand and commission access audits;
- create and manage an access improvement programme;
- maintain accessibility in buildings and working practices; and
- respond effectively to the legal requirements of the Disability Discrimination Act 1995, with particular reference to those parts of the Act that come into force in 2004.

The section covering access auditing gives information on why audits and appraisals should be carried out, explains the audit process and how it fits into an access improvement programme. The section on access management covers the implementation of improvements and the importance of ongoing access management to ensure accessibility is sustained in use.

The manual also gives guidance on handover and commissioning of new and improved buildings, feedback procedures, post-occupancy evaluation and ongoing management of the accessible environment. Relevant legislation and standards are described, explaining the effect on accessible design and giving information on duties and obligations.

The design criteria cover access to and use of buildings and take account of the needs of a wide range of users. The design guidance can be used when designing new buildings or taken as a standard to assess and improve existing ones.

Appendix A contains a number of sheets of 'general acceptability criteria', which can be used to highlight where access problems exist. Extracts from sample audit reports are included in Appendix B to illustrate various report formats. Appendix C gives sources of reference and further information.

Throughout the manual there are boxes giving hints and tips. The information given in the boxes covers issues that are not always found in standards, legislation or other guidance, and includes advice and thoughts that come from the authors' experience.

Inclusive design is about people and their needs, and, in the context of this manual, how these relate to the design, use and management of the built environment. The manual includes comprehensive information on standards, legislation and good practice, but also recognises that to achieve a truly accessible environment designers and operators of buildings must move beyond compliance with standards and adopt a new way of thinking. Taking a creative approach, considering the needs of everyone, integrating those needs into good, thoughtful designs and practices will help achieve an accessible, inclusive, built environment that enables people to participate fully in all aspects of society.

1 Access audits and appraisals

Introduction

The purpose of an access audit is to establish how well a particular building or environment performs in terms of access and ease of use by a wide range of potential users, including people with disabilities, and to recommend access improvements. The process involves a thorough site inspection, an assessment of the management and use of the building or environment and the preparation of a report that identifies accessible user-friendly features as well as access problems. The report should recommend access improvements, prioritise action and indicate where improvements can be made through the building's maintenance programme and by altering management procedures. The audit is the first part of the access improvement process and should be followed by the preparation of an access plan setting out the strategy for the implementation of the proposals, and leading to ongoing management of the accessible environment. This will enable building owners to plan ahead for costly improvements and allow alterations to be made cost effectively and over time.

Usually, an audit will consider the needs of all users, and potential users, of a building or environment and assess the factors affecting independent use and access to services. However, an audit may be carried out in response to a particular issue, such as how to meet the needs of a disabled employee in the workplace, and this may affect the scope of the audit and the standards used in assessing accessibility.

It is important to involve the building owners, managers and operators, as appropriate, in the audit process. Many of the issues that arise will be concerned with the operation and management of an environment, not just the fabric.

The term access appraisal is used to describe an audit of the proposals for a development. This involves making a detailed assessment of the proposed level of accessibility in a building or environment using drawings, specifications and consultation with the architect or designer. To be most effective an appraisal should be ongoing throughout the design process.

An audit can also be carried out to assess service provision. This may be useful for an organisation that is defined as a service provider by Part III of the Disability Discrimination Act 1995 (DDA) or that wishes to improve its services generally.

Why carry out an access audit?

An access audit will give a picture of the level of accessibility in a building, identify points of good or bad access, identify areas of need that are not catered for and is a first step in the process of improving accessibility.

> The increase in accessibility that can result from an access audit and subsequent access improvements will benefit all users of the building. Issues such as poor signs, doors that are heavy to open and lack of handrails affect everyone, not just people with disabilities.

The reasons for carrying out or commissioning audits and appraisals may include:

- Disability Discrimination Act 1995;
- funding conditions – Lottery funds, grant from trust or other body with specific access requirements;
- to gather data on buildings for comparison or analysis;

- to check compliance with certain standards, such as Part M of the Building Regulations;
- company policy on equal opportunities;
- public relations/company image;
- conservation by use of historic buildings;
- pressure from lobby groups;
- awareness of a particular problem.

The Code of Practice relating to Part III of the DDA clearly suggests that service providers are more likely to be able to comply with their duty to make adjustments in relation to physical features if they arrange for an access audit of their premises to be conducted and draw up an access plan or strategy. It states that 'acting on the results of such an evaluation may reduce the likelihood of legal claims against the service provider' (DRC 2002).

Figure 1.1 A good level of accessibility can benefit a wide range of people, including people with disabilities.

Audit preparation

Information about the building, the occupier, the services provided, the length of time the building will be occupied, the available budget and future plans for alterations or refurbishment should be collected before the actual audit commences. Taking all these factors into account will help ensure that the audit covers all the necessary issues and that the recommendations made are relevant, practical and likely to be effective.

Information on the commissioning and the scope of the audit should also be confirmed prior to commencement.

Information about the building

The size, number and location of buildings should be confirmed, along with their age and type and use. There will be particular issues relevant to specific building types, for example, an education building may have lecture theatres or laboratories with particular requirements, and a theatre will have particular acoustic requirements. The location of public transport and car parking should also be considered.

Historic buildings   Whether a building is of any special architectural or historic interest is also relevant, especially if it is listed or there are restrictions on alterations. It should not be assumed that listed buildings cannot be altered. Guidance issued by English Heritage 'Easy Access to Historic Properties' (EH 1995) acknowledges that in certain circumstances alterations can be acceptable if part of a long-term strategy for use, though recommends that they should be reversible wherever possible. The guidance quotes Planning policy guidance note; planning and the historic environment (PPG 15), which states:

'It is important in principle that disabled people should have dignified easy access to and within historic buildings. If it is treated as part of an integrated review of access requirements for all visitors or users, and a flexible and pragmatic approach is taken, it should normally be possible to plan suitable access for disabled people without compromising a building's special interest.'

When considering alterations to an historic or listed building, it is important to establish the extent of the listing. In some buildings, it may be only particular features, such as a façade or staircase, that are listed and alterations may be possible to other parts of the building. Some alterations may be possible even to listed parts if no permanent damage is done to the historic fabric. For example, it may be possible to fit a wheelchair stair lift as long as it can be removed at a later date with no damage to the fabric having occurred. Consultation with the local planning or historic buildings officer is always a good idea.

Even if physical alterations are not possible, there may be other ways of getting around an inaccessible feature, or providing the service or employment opportunity being offered within the building by a 'reasonable alternative method'.

Figure 1.2 Access to historic buildings can be improved with sensitive, well-designed alterations.

Future plans   The length of future occupancy may influence proposals for improvement; whether the building is owned or let is also relevant. Plans for refurbishment or alterations should be taken into account as they may affect access or they may present an opportunity to make access improvements.

Information about staff, management and building use

It is essential to have information about the occupier and the nature of his or her business, for example, whether the occupier is an organisation that offers, or could offer, a service to members of the public. How a building or environment is used and managed can have a huge impact on its accessibility. A well-designed, accessible building can be made difficult or impossible to use unless management and maintenance procedures take account of access issues. Information on policies, procedures and building use should be collected and relevant issues identified.

Contact with staff, including any employees or other building users with disabilities, will give an opportunity to discuss how the building is used and whether there are any specific access problems. Questions to be asked could include:

- Have staff or management had any access training, or is any planned?
- Is there a member of staff in the company with responsibility for access issues?
- Is there a human resources department or a health and safety section in the company?

Costs and benefits

In some cases the available budget for improvements to the building may also be relevant as this can affect the scope of any alterations that may be recommended. For example, if the size of the budget prevents the installation of a lift, this may well affect the recommendations for using an upper floor for providing a service or as a place of employment for a current or potential disabled employee.

In certain situations, an employer may be able to obtain grants under schemes such as 'Access to Work', which may affect the extent to which the employer can carry out physical alterations to a building or the provision of facilities for an employee. Knowledge of such schemes and other potential sources of funding are an important part of the service offered by an access auditor or access consultant.

VAT should also be considered. At the time of writing some goods and services can be zero-rated when supplied to charities or for the use of specific disabled people in private residences. It should be made clear whether or not VAT and other costs, such as fees, are included in cost advice given as part of an audit.

The most expensive access improvements may not benefit the greatest numbers of people. Often smaller alterations will have a more wide-ranging effect.

Access information

The scope of the report and the standards against which access will be assessed should be confirmed prior to the audit. Matters that should be checked include the following:

- the standard against which the building is to be assessed;
- whether the needs of staff are to be considered as well as customers and visitors;
- the access policy of the organisation;
- particular access problems in the building to be audited.

It can be useful to check whether there has been previous contact with an access group, or an access officer, and whether an access audit has been carried out in the past.

Egress, especially in an emergency, is a critical part of the accessibility of a building and is usually affected by both the design of the building and how it is managed. If emergency egress is to be considered, both these issues will need to be audited; if it is not included, this must be made clear to ensure no misunderstandings occur as to the extent of the audit.

Commissioning

Information concerning the commissioning of the audit is also critical. Instructions covering fees, timescale and scope should be confirmed in writing. Issues of confidentiality should be discussed, together with determining whether or not photographs can be taken during the audit and whether the auditor needs to be accompanied around the building. Wherever possible, copies of floor plans of the building being audited should be obtained as well as copies of visitor information and publicity material.

> In larger commercial and public buildings, floor plans may be displayed alongside fire certificates. These can give an indication of the size and layout of the building.

It is useful to consider whom the report is for and how the information will be used. If the audit information is to be transferred to an internal maintenance or improvement programme, the use of compatible formats or software when collecting the information will assist this process.

> Photographs should be included in the report and can also be a valuable record for the auditor. Factors affecting accessibility, such as lighting, signage or obstructions on circulation routes, can change on a daily basis. It is important to have some record of the level of accessibility at the time of the audit, if only to be used if a dispute arises at a later date.

How environments are used

It is important to consider the ways in which each area of a building or environment is used, managed and operated rather than just relying on the generic title to describe the function. A building such as a shopping centre or hospital will contain many areas with different functions, where the physical design and type of use of each area may affect the access requirements. For example, within a hospital, the factors affecting accessibility will vary considerably across circulation routes, refreshment areas, wards, operating theatres, waiting rooms and consulting rooms. Some areas will be used by members of the public who are unfamiliar with the building and whose individual needs will be unknown, whilst other areas will be occupied by staff whose needs can be assessed and met. In some buildings visitors may have to rely upon signs for information, in others there may be reception desks where staff are able to identify particular user needs and offer assistance where required. Access may be restricted to some parts of buildings, whereas other areas may be open to everyone.

Entire buildings or parts of buildings can be classified according to use and this approach can be helpful in understanding how to provide services in a non-discriminatory way and how to improve accessibility in ways that will suit the needs of all users.

Use classification   There are four use classifications described here:

- use classification 1 – complete freedom of movement;
- use classification 2 – controlled entry/freedom of movement;
- use classification 3 – free entry/controlled movement;
- use classification 4 – controlled entry/controlled movement.

Use classification 1 – complete freedom of movement   A building or area in this classification would be one where the user or visitor is free to enter, wander around, probably in no particular sequence, and leave without the need to make any contact with potential assistance points such as a reception/information desk or security point. Environments that fall into this category may include shops and shopping centres, department stores, supermarkets, some hospitals and non-fee paying museums and exhibitions.

An environment in this use classification is likely to contain long travel distances. The provision of seating, preferably where visitors do not have to pay to sit down as in a café, will be helpful to many people, especially older people and those with disabilities.

In environments that allow users complete freedom of movement, information will need to be provided remotely, usually by signs. The provision of an information point or help desk may assist some visitors but will not remove the need to provide information remotely, as it cannot be assumed that all visitors will make use of it.

Environments which allow the type of freedom of movement described here are likely to be the most difficult to design and manage to ensure appropriate levels of accessibility for all users. There may be a wide range of needs to be met and little opportunity to provide specific assistance. In such environments the provision of good environmental services (lighting, acoustics, colour and luminance contrast, etc.), appropriate communication facilities (signs, audible and visual information systems, etc.) and ongoing staff training are essential.

Many people, but especially older people and those with hearing impairments, dislike environments with poor acoustical qualities, or ones that contain equipment giving a high level of background noise.

Use classification 2 – controlled entry/freedom of movement
Environments in this category will have some point of control, usually a payment desk or security point, but, after passing that point, users will be allowed the type of free, usually unrestricted, movement described above. This category might include sports halls, ice rinks, fee-paying museums and art galleries, some hospitals, exhibition halls, libraries, some educational buildings, some hotels and some offices.

> Good lighting is essential at any reception desk or information desk to ensure that people with sensory impairments are able to communicate effectively. However, it is not just more lighting that is needed, it is appropriately designed lighting that maximises visibility without increasing glare or shadows.

In such environments there is still a need to deliver information and assistance remotely once the visitor has passed the point of control, but there is also the opportunity, if management practices are appropriate, for particular user needs to be identified and for assistance to be offered. It is important to consider the potential needs of visitors, have the ability to provide appropriate assistance if required, and have contingency plans in place for those situations which are more difficult to foresee.

Use classification 3 – freedom of entry/controlled movement
This type of environment will usually have a central entrance through which the visitors enter, but once inside, movement around the building will be restricted. Examples of this type of classification include town halls, civic centres, major post offices, law courts, airports, bus and railway stations, some hospitals, theatres, cinemas, some multi-tenanted offices, restaurants, banks (as a customer) and some hotels.

Some larger buildings may have areas of controlled movement combined with other areas where there is complete freedom of movement. An example is a hospital where there may be complete freedom of movement around the refreshment areas, shops and corridors, and controlled entry to wards and clinics. The methods of providing information may vary according to the type of use.

Use classification 4 – controlled entry/controlled movement   In this type of environment, security will usually be a major issue and the type of visitor will be restricted. Controlling the entry to an environment allows an assessment of the needs of all users to be made at the initial point of contact, providing that that initial means of requesting entry to the environment is, itself, fully accessible.

Examples of this type of classification would include offices with car park controls or entry phones/CCTV, some schools, some multi-tenanted offices, research laboratories, and some banks (as a customer or an employee). In buildings in this type, management issues will be of great importance. Consideration should also be given to how the building will actually function during all the time it is in use. For example, a school may fit into this classification during the day but may have complete freedom of movement in the evening if it is being used for evening classes or community projects.

> Signs provided in schools should be clear and simple and designed to meet the needs of the children using the building during the day, but also the adults who may use the building in the evening.

An analysis of a building using this system of classification can help to give a more accurate picture of the specific access requirements of each individual area and can highlight opportunities for improvements.

Who uses the environment?

It is also important to consider who uses the environment being audited and how this may affect general and specific recommendations to improve accessibility. A service provider has to anticipate the needs of his or her customers, but an employer can identify the needs of his or her workforce, though not those of potential employees. Whether or not specific access requirements can be identified will affect the advice given in an audit.

For example, an employee may require a greater amount of circulation space at a workspace, a particular desk arrangement or the provision of special equipment. These measures will be identified in relation to a current or future employee and so can be carried out when the need arises. In addition to recommendations aimed at achieving a good overall standard of accessibility, an audit report should recommend that procedures are in place to deal with potential future needs that cannot be identified in advance, and identify issues that are not currently causing problems but may need to be dealt with in the future.

Recommendations to alter storage arrangements in a workplace to accommodate reach from a wheelchair may not be necessary where there are no wheelchair users using the storage facilities. However, the access audit could raise the issue and note that alterations should be made when the need arises.

Who should carry out an access audit?

To carry out an audit successfully the auditor or consultant needs to have two main areas of knowledge:

- a comprehensive knowledge of the needs of disabled people in order to identify and understand the difficulties caused by the built environment;
- a working knowledge of building construction and some knowledge of design to be able to identify practical and appropriate solutions.

An auditor does not require the ability to produce detailed designs, though some may have this skill. An audit will usually stop short of design solutions, though may include sketches showing, for example, how a change of level could be overcome with a lift or ramp.

Also helpful are skills in surveying, assessing and communicating views and opinions, especially in report format.

There is a view that an access auditor should be a disabled person, as only then will he or she have first-hand knowledge of the needs of disabled people. Others say that disabled people are individuals and it should not be assumed that they will have an in-depth knowledge of the access needs of others. All disabled people are likely to have experienced problems with the built environment, and this may lead to an awareness of access issues, but a wheelchair user will not necessarily be aware of the specific needs of a person with a visual or hearing impairment or vice versa.

However, it is clearly desirable to involve disabled people in the audit process and this can be done in various ways. Where there is a local access group, members could be invited to contribute to the audit process. Members of staff who are disabled must be consulted as they will be aware of access issues in the building and also may have particular access needs. On larger projects it may be possible to set up a focus group or committee of people representing different interests to provide input into the audit or appraisal process.

If an access group is involved or if individual disabled people are invited to join a focus group it will be necessary to allow for additional costs incurred. It should never be assumed that people will contribute their expertise without payment.

The audit

When to audit

An audit should take place when a building is occupied, as this gives an opportunity to observe procedures and assess the building in use. If a building is busy at certain periods of the day, or perhaps used for a different purpose in the evening, the audit should include assessment of this use also.

The time of day and the weather may affect any observations and should be noted in the report. Comments on glare, colour contrast and general illumination of the environment will relate precisely to conditions at the time of the audit and these may vary from day to day or even from hour to hour. In addition, rain, leaves, snow and ice can also affect the safety of external routes and raise issues of maintenance. Just because certain factors are not present at the time of the audit does not mean their influence on the accessibility of an environment can be ignored.

> Always think about solutions while auditing. It is often easier to identify a possible solution when in the building actually looking at the problem.

What to audit

The audit should cover both physical features and issues of management and use.

Physical features usually include:

- **external environment** – including approach, parking, transport links, routes, street furniture and external ramps and steps;
- **entrance** – including visibility, entry controls, doors, thresholds and lobbies;
- **reception area** – including layout, reception desk, waiting area, signs, visual and acoustic factors;
- **horizontal circulation** – including ease of navigation, corridors, doors, directional information, internal surfaces;
- **vertical circulation** – including internal steps and stairs, ramps, escalators and lifts;
- **WCs** – general provision, WCs for ambulant disabled people, accessible WCs and baby changing facilities;
- **specific facilities** – such as changing areas, bathrooms and showers, bedrooms, storage, refreshment areas, service desks, waiting areas and assembly areas;
- **controls and equipment** – coin and card operated devices, building services controls, window controls, alarms, entry phones;
- **communication systems** – such as telephones/text phones, lift voice announcers and audiovisual displays;

- **emergency egress** – including escape routes, refuges, alarms, fire protected lifts, emergency lighting;
- **signs and way finding** – including overall layout of building, sign type and location, use of landmark features, maps and guides, colour and contrast, audible features and olfactory features;
- **lighting** – general and workplace;
- **acoustic environment** – including background noise, hearing enhancement systems, acoustic conditions suitable for intended use.

It is not just the physical features of a building or environment that affect accessibility. Access to information, staff attitudes and working practices can have a major impact on actual and perceived accessibility and can also sometimes provide the easiest and most cost-effective ways to improve accessibility.

For many disabled people, access to information is as important as physical access. The availability of publicity and information material is essential to allow disabled people to participate, contribute and make planned use of services. Information should be provided in a variety of formats to be appropriate to people's needs. Staff should be properly trained to be aware of the needs of disabled people and to manage the environment and the services in a way that allows equal access to all users.

The audit should consider how such matters are dealt with and include information on:

- **access to information** – publicity and information material, formats available, text phones, hearing enhancement;
- **attitudes of staff and management** – training in disability and accessibility issues;
- **management practices, policies and procedures** – including those relating to emergency egress;
- **maintenance issues** – in relationship to achieving or maintaining accessibility;
- **use of the building** – the way in which the building or environment is used by employees, visitors and members of the public.

To give a comprehensive assessment of the level of accessibility and to identify the opportunities for improvement an audit must take account of these issues.

Access improvements can often be made at no cost by increasing awareness and altering policies and procedures.

Assessment standards and dimensions

The standard for assessment should be agreed before the audit commences. It might be general good practice, or a particular publication or standard could be identified, such as BS 8300:2001. The audit should always consider usability, and look at how people move through the environment and use the facilities safely and as independently as possible.

The skill of the auditor or access consultant is in judging what is reasonable; best practice standards may not be found in existing buildings and may be impossible to achieve.

It is important to take dimensions on site and to be clear what standards are being used. It is unhelpful to note that a feature is adequate if the standard of assessment is not clear. The standards adopted for the audit should be used consistently throughout the building.

The audit should include detailed factual description, with relevant dimensions as appropriate, and fully describe physical features, facilities and management practices.

Audit equipment

The auditor will need, at the very least, a tape measure and some means of recording the information collected. It is desirable to use, in addition, a camera or video camera and, where appropriate, a door weight measure, a light meter, a sound meter, a gradient measure and a means of testing an induction loop.

> Where equipment such as light meters and weight gauges are not available, subjective judgements may have to suffice. It is useful to have these judgements made by people who will be most affected by the feature being assessed, for example, by asking a wheelchair user's opinion of a deep pile carpet or by taking the advice of a person of reduced strength or mobility on door opening and closing forces.

General acceptability criteria

Lists of 'general acceptability criteria' are given in Appendix A. These lists, relating to car parking, external areas, entrances, stairs and steps, ramps, lifts, horizontal circulation and accessible WCs, can be used to highlight where access problems exist. The lists should not to be used as a substitute for an audit, but as part of an initial stage to identify where standards fall short of good practice and further investigation is necessary.

The journey

The audit should take the form of a sequential journey, starting at the point at which people may start their visit to the building, continuing into and around the building, and finishing with safe egress from it. It is useful to separate out some major elements, such as vertical and horizontal circulation, WCs, signs, lighting, auditoria and means of escape, to help structure the audit. Plans of the building can be used to organise a route for the audit, check off the areas visited and annotate with the information collected.

> Standard checklists can be useful, as long as the checklist does not limit the information collected. A tick in a box is very unlikely to give sufficient information to fully describe an issue.

Recording the information

Information can be recorded using notes, sketches, Dictaphone (or dictation to an assistant), camera, video camera and annotation of existing drawings. If a laptop or palm top computer is used, the information can be put into a predetermined format or onto appropriate software.

> It is useful to record examples of good access as well as access problems. Issues not covered in design guidance may arise on site or from discussions with staff and new good practice may be learnt.

Audits of services

The concept of the service audit has come into being largely because of the DDA. The Act is concerned primarily with employment and the provision of services, not the design of buildings. When assessing an environment to see if the requirements of Part III of the DDA are met, it is the provision of services that is the critical issue. A service audit differs from an access audit as the focus of the process is on customer service, rather than the built environment and the effect it has on the provision of services.

The assessment is of the services that are provided by an organisation to its clientele and, in relation to the DDA, the aim would be to identify any areas of service provision that are discriminatory. The measurements taken are of service provision, attitudes, training, etc., and the information is gathered from staff and recipients of the service.

The methods used to gather information can include assessment of staff education and training, staff motivation and commitment and customer and staff feedback. Face-to-face interviews with staff and customers, focus groups, telephone interviews, written questionnaires and mystery shopper techniques can also be used. Information is collected and analysed and used to produce an objective assessment of the level of service provided, highlighting opportunities for improvement.

As with an access audit, there is a need to set out assessment criteria, identify good service and make recommendations for improvements, with costs and priorities where appropriate.

The report

Preparing the report

In preparing an audit report, it is essential to consider the purpose of the audit and for whom it is designed. It is critical that the information is presented in a way that will allow the building manager or owner to make best use of it. The audit report, and its format, should encourage, not hinder, implementation of the proposed improvements.

The purpose of the report is:

- to explain how the building or environment can best cater for the needs of all its users;
- to set out priorities and procedures for carrying out the recommended improvements;
- to identify the availability of resources, where appropriate;
- to set out how accessibility can be sustained in use.

The report should describe the building or environment and the current access situation, recommend access improvements, prioritise action, give costs and indicate where improvements can be made through a maintenance programme or by management action.

Information can be presented using:

- the narrative method, with a description of the existing situation and identification of the problem, followed by a recommendation for remedial action;
- the tabular method, where as much information as possible is given in tables with a minimum of text;
- a combination of both, such as a narrative format with a tabular summary.

Examples of audits and various formats are given in Appendix B. However, whichever format is used, a standard report should always include the following:

- an introduction describing the function of the building and how the building is currently being used, noting any particular access problems;
- a detailed description of the existing access situation, with dimensional information and considering each identified element of the audit in turn;
- recommendations for improvement, using a stated standard or standards;
- priority ratings for the recommended improvements;
- cost information where required;
- summary, identifying the main points of the report.

A report should include photographs and plans of the building, and sources of further information where required. It can be useful, particularly where there are a number of buildings being assessed, to include a manual of design guidance as an appendix to the report. See example 1 in Appendix B.

Description of current situation

There should be detailed information collected and presented describing the elements of the building and current access arrangements. It is not sufficient to state that an element is inadequate or incorrect; dimensions should be given with a detailed description.

It is also helpful to note points of good access, not just elements that fall below a certain standard.

Audits which only highlight or identify problems can be very demoralising. Wherever possible, highlight some areas of good practice, if only to ensure that the service provider or employer does not simply give up and do nothing!

Recommendations

Recommendations should explain how a feature could be altered or removed to improve access. Recommendations will cover many areas and not all will require alterations to the building. There may be a management procedure that could be improved, such as better policing of disabled parking bays.

Many issues will be able to be dealt with as part of a regular maintenance programme; the introduction of colour contrast in redecorations is an example. Relocating a service or facility to another building could also solve an access problem.

Refurbishment of a building may give opportunities to make access improvements such as altering door widths or automating door opening. It may also be possible to adjust floor levels or incorporate ramps.

Priority rating

It is useful to prioritise recommendations. An example of a rating system is as follows:

(1)   implement immediately to eliminate a severe barrier or a hazard to access to and use of the building by disabled staff or visitors, including potential health and safety or occupier liability issues;

(2)   implement as soon as practicable to improve access;

(3)   plan action to be implemented when relevant area/element of building is refurbished/upgraded;

(4)   plan adaptation to be part of a workplace assessment to be implemented to suit the needs of an identified member of staff;

(M) implement and regularly review as part of specific or regular maintenance or renewal;

(A)   no action is reasonably practicable. Arrange for assistance to be readily available or for the service or employment opportunity to be provided by an appropriate and reasonable alternative means.

The rating given to recommendations will depend upon a number of factors and may vary from project to project. It may also vary for identical access solutions for different service providers or employers. Improvement to accessibility will be a major factor, but priority may also be influenced by the availability of funds for certain types of work, proposed works to the building or length of future occupancy. A method for setting priorities must be agreed in advance of the audit.

Costings

Information on the cost of the recommended improvements is a useful element of an access audit report. The source of the cost information, such as the 'Access Audit Price Guide' (BCIS 2002), should also be given.

In many cases it will not be possible to provide accurate and detailed costings unless the recommended improvements are taken to detailed design stage. However, it will usually be possible to provide broad band cost information.

Recommendations can be categorised as low, medium or high cost or cost bands can be given, say:

- up to £500;
- £500 to £5000;
- £5000 to £10 000;
- £10 000 to £20 000;
- over £20 000.

The range of each band will depend, to some degree, upon the type of recommendations and the overall budget.

As with all other maintenance or improvement work, the costs will vary according to many factors, such as whether in-house staff or contract staff are used to undertake the work.

Always establish this before offering cost bands or clearly identify the ground on which costs have been developed.

Summary

An executive summary is a useful device and gives the opportunity to set out the main points in a report succinctly. The summary can provide a picture of the current access situation in a building and identify the key issues in the report.

> Often the summary will be the first part of the report that is read, if an overview of the issues is sought. Too much care cannot be taken to ensure that the summary is well constructed, concise and accurate.

The summary should also draw attention to issues of building operation and procedure and set out how the audit fits in to the access improvement process. It is critical that the audit is seen in context and as the first part of a process, not the conclusion.

Access appraisals

An access appraisal is an assessment of the access provision in proposals for a new development, refurbishment or alteration. Appraisals can also used to assess access policies, access briefs, publicity and information material and maintenance programmes. Appraisals are carried out for many of the same reasons as audits, and the information that should be gathered beforehand is similar.

When looking at proposals for buildings and environments the appraisal follows the pattern of an audit, but the journey is taken on paper and in discussions, rather than through the building. Potential access problems can be identified or anticipated, however, a comprehensive review depends on there being a sufficient level of detail and information available. Physical features should be assessed, as in an audit, but the guidance given should also include issues of management and use.

Recommendations for access improvements should be given, though in some cases these may not be as prescriptive as when carrying out an access audit. The appraisal may suggest that a particular layout or feature will make access difficult and recommend that an area be redesigned bearing in mind certain factors. Usually, an appraisal will not include detailed redesign of a proposal, but will provide the necessary information to allow the designer to improve the proposed accessibility. It is preferable that the auditor or access consultant is involved in a project through all stages of design, construction and commissioning.

Information available

The level of information provided in proposals will vary depending on the stage the design process has reached and this will affect the guidance that can be given. Often detailed information will not be available when the appraisal is carried out and requests should be made for further detail. As well as looking at plans of the proposals, specifications and schedules should be checked. Inevitably there will be some areas where no information is available, but detailed recommendations can still be provided to ensure that issues are picked up at the appropriate time. An access appraisal report can become a checklist for a designer to use as the design develops, to ensure that all the relevant areas are covered.

The report

An access appraisal report can have a similar format to an audit report, listing current features and making recommendations for access improvements. It may not be appropriate to include priorities and costings if the appraisal is carried out at design stage. However, as an appraisal is usually not a static exercise, as the design will be evolving during the appraisal process, it may be more appropriate for the appraisal to take the form of a series of 'plan checks'. These can be carried out at various stages and make recommendations that can be 'ticked off' once implemented.

Ongoing involvement of auditor

If an appraisal is requested for a funding or grant submission, it may take place at an early stage in the design process and the involvement of the auditor may be limited. It is preferable for an auditor to have an ongoing relationship with the designer throughout a project, as this will allow assessment of more detailed proposals and specifications as they become available.

It is becoming more usual for an access consultant to be involved in large projects from the start, as happens with consultants in other fields. This allows the consultant to have input throughout the design and construction process and this level of involvement is likely to result in a more integrated design solution. There should also be a method of passing on access information once the building is commissioned to ensure that accessible features are understood, made use of and not altered unknowingly.

Building use and management

Issues of building use and management need to be taken into account in an appraisal, as in an audit. Information relating to these issues should be included in any report and areas where use and management may influence access should be highlighted.

Skills required

Skills in reading and understanding plans are required to carry out an access appraisal of proposals using drawings. These include the use of scale, drawing conventions and symbols. There will be a need to translate the information given on the drawings and in the specifications into a three-dimensional picture of the building or environment to be able to anticipate fully any potential access issues.

As with access auditing, a knowledge of the environmental needs of disabled people is required, and a working knowledge of building construction.

2 Access management

Introduction

An access audit is the first step in improving the accessibility of a building or environment. On completion of the audit, the access improvement process, to be effective, should continue with the preparation of an access plan. The access plan provides a strategy for the implementation of the improvements, leading to ongoing management of the accessible environment.

Active management of the improvement process can help ensure that the work is done cost-effectively, incorporated into existing maintenance procedures or refurbishment programmes where appropriate and will allow planning for future maintenance or management input. Accessibility does not end with specification, it needs to be sustained in use.

Ongoing access management

The access plan

An access plan is a strategy for improving accessibility developed from an access audit and can help ensure that access is an ongoing concern, help identify opportunities for change and is a useful tool for meeting obligations under the DDA.

The Code of Practice relating to Part III of the DDA states:

'Service providers are more likely to comply with their duty to make adjustments under the Act if they:

- audit physical and non-physical barriers to access for disabled people;
- make adjustments and put them in place;
- provide training for staff which is relevant to the adjustments;
- draw the adjustments to the attention of disabled people;
- let disabled people know how to request assistance;
- regularly review the effectiveness of adjustments and act on findings of the review.'

An access plan can incorporate the above recommendations and start to ensure that access is an ongoing issue that is included in, for example, planned maintenance programmes. The adoption of a strategic approach will allow access improvements to be made in a sensible and cost-effective way, where major items of expenditure can be identified and budgeted for and other items included in everyday management or maintenance.

The access plan should incorporate:

- a schedule of works that has been devised to take into account the priority and cost information in the audit;
- processes to allow regular updating of the audit information;
- links to maintenance and management procedures.

The plan should set out procedures to ensure that when opportunities for access improvements arise they are recognised and not lost and should identify where maintenance or management input is required to make or sustain access improvements.

Issues requiring management input

The following management and maintenance issues should be considered to ensure that access is achieved and maintained:

- **car parking** – allocation and management of designated bays;

It is not sufficient to provide designated car parking bays. There must be an appropriate and effective management procedure in operation to ensure bays are only used by disabled people. If the use of bays is not controlled this may well be seen as a discriminatory management practice.

- **external routes** – keeping in good repair and free of obstructions and leaves, ice, snow and surface water;
- **doors** – adjustment of door closers, ironmongery kept in good working order, pass doors kept open;
- **horizontal circulation** – keeping routes free from obstructions, keeping furniture layouts and seating arrangements accessible, making available auxiliary aids such as temporary ramps;

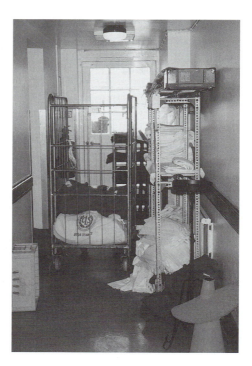

Figure 2.1 Obstructions such as these should be identified as problems associated with working practices. Assessing if it is a space problem or a staff training issue should be part of any access audit.

- **vertical circulation** – regular checking of lifts to ensure floor of car aligns with finished floor level, maintenance checks on short rise and wheelchair stair lifts;
- **WCs** – checking manoeuvring space in accessible compartments not obstructed by bins, sanitary disposal equipment, etc., replenishment of toilet paper and paper towels in accessible WCs as well as other WCs;
- **communication** – new signs to integrate with existing sign system, no *ad hoc* homemade signs, and all information kept up to date, signers and audio description services provided as necessary, appropriate provision of accurate access information and other literature;
- **hearing enhancement systems** – advertising, regularly checking and maintaining, ensuring loop positions at counters are available to users;
- **alarm systems** – checking and staff training in procedures;

> If vibrating personal pagers are used to alert people who are deaf or hard of hearing when the fire alarm is activated, it is essential to ensure that batteries are kept charged, regular maintenance on the pagers is carried out, and that they are compatible with the systems they are linked to – at all times.

- **surfaces** – ensuring cleaning does not cause slippery surfaces, maintaining junctions to avoid worn surfaces becoming tripping hazards, replacing like with like, maintaining colour contrast in redecoration;
- **lighting** – replacing of bulbs, keeping windows and light fittings clean;
- **means of escape** – staff training, regular practices, maintenance of fittings and equipment, reviewing evacuation procedures;
- **security** – ensuring security procedures do not conflict with accessibility good practice;
- **training** – ensuring staff training is ongoing and appropriate;
- **health and safety policies** – to include information on access, risk assessment;

- **responsibilities for access** – identifying who is responsible and gives approval for improvements, setting priorities, ensuring access is included in maintenance and refurbishment programmes, arranging temporary relocation of services provided within building where there are access limitations, providing auxiliary aids, reviewing numbers of disabled people using a service, establishing and running user groups;
- **funding for access improvements** – identifying specific access funds or grants, funds for specific employees such as 'Access to work', use of the maintenance budget;
- **policy review** – regularly reviewing all policies, practices and procedures affecting access.

There may well be other issues, relating to specific buildings types and functions. The access plan should take account of any particular requirements.

Opportunities for improving accessibility

Within the general maintenance of any building and environment there are often opportunities to introduce measures that improve accessibility, most of which can be undertaken at little or no extra cost.

For example, throughout the life of most buildings there will be repair and refurbishment carried out, perhaps in response to:

- schedules of dilapidations at the end of a lease;
- work undertaken to meet ongoing health and safety or environmental health requirements, such as redecorating or replacing tiling in toilets and kitchens or upgrading lighting;
- replacement of finishes as part of ongoing repairs and refurbishments, including the renewal of floor finishes or redecoration;
- the replacement of old or broken facilities such as taps, light switches, furniture, toilet flushes and larger items such as lifts;
- change of corporate identity, offering opportunities to improve company information and signs.

Figure 2.2 Floor finishes such as this, poor in terms of pattern and maintenance, can be replaced as part of ongoing refurbishment. However, management practices to assist people who have difficulties in using such a floor will be necessary until that occurs.

Alterations and extensions may not be required to be accessible by building regulations, but can afford opportunities for improvement. If external areas are refurbished, there may be opportunities for repaving in different materials to identify routes or adding outdoor seating. Inside the building there may be opportunities for improvements during redecoration programmes to change wall tiles in WCs or introduce new colour schemes to provide contrast. Major alterations may give the chance of levelling out differences in floor levels or providing ramps.

Elements of the built environment, such as buildings, pedestrian areas and transport infrastructure, are with us for a long time – but their life is dynamic, not static. There are often opportunities to improve accessibility for everyone – but there are also a greater number of opportunities to make it worse if all the relevant issues are not fully understood.

Opportunities to maintain accessibility

The adoption of good management practices such as planned maintenance programmes offers considerable opportunity to address and improve accessibility. The access audit can highlight maintenance issues that affect accessibility, and a planned maintenance programme can ensure that there is a continuing commitment to maintain accessibility where it has been achieved. It can involve regular inspection routines to ensure all aspects of the building are kept up to a good standard of repair and allow planning of expenditure for maintenance of, and implementation of, improvements.

For example, even something as simple as painting nosings onto a staircase where they did not exist previously will require the nosings to be reviewed on a regular basis, and for money to be available to repaint the nosings when required. The installation of induction loops requires an ongoing commitment to maintenance

Figure 2.3 Once a nosing is painted, a commitment is made to re-inspection, repair or redecoration/replacement. The implications in terms of duties and obligations under the DDA should be considered for all access issues.

and testing; a platform lift may require a rather more expensive maintenance contract.

Issues of maintenance that arise in the access audit should be identified and linked to maintenance programmes to ensure that they are dealt with on an ongoing basis.

It is not sufficient to simply provide a facility such as an induction loop. The loop must be operational when the person who needs to use it actually visits the building.
If it is not functioning, they could argue that management procedures covering the testing and maintenance of equipment are discriminatory.

A lift that is not working when it is required will discriminate against those who need to use it to access a service or employment opportunity. Lifts can break down and the service provider or employer may be required to answer questions such as:

- What procedures are in place to ensure problems with such an important access facility are dealt with efficiently?

and, if it happens regularly,

- What has been done, in terms of either management practices, policies and procedures, to limit the impact of the problem for disabled people?

Good provision is essential – appropriate ongoing management is critical.

Occupying the building

When a new building is occupied, or after improvements have been made to an existing building, there will be a number of opportunities to ensure accessibility is fully achieved and reviewed.

Handover and commissioning of new or improved buildings

At the handover and commissioning stage of a new development, or after a programme of access improvement works has been carried out, there may be some small alterations and adjustments that will benefit accessibility that were not obvious at design or construction stage. Issues such as the location of signs, or the redirection of lighting to avoid glare, need to be dealt with to ensure that the best possible level of accessibility is achieved. There will also be written procedures to prepare or revise, such as health and safety policies, employment policies and security procedures, all of which should take account of access issues.

Feedback

Information should be passed back to those responsible for the design or improvement programme on whether certain aspects or improvements are successful. Feedback can prevent faulty details being used again and inform designers, clients and access officers of how their ideas work in practice. A formal system of collecting and using feedback to inform future projects can prevent some very useful information being lost forever. An accessibility review should be carried out on all projects and records kept of improvements and results.

Post-occupancy evaluation

Post-occupancy evaluations look at such issues as:

- the effectiveness of improvements;
- whether more disabled people are using the building or service;
- whether management is making the most of the improvements;
- whether the improvements have been effectively publicised.

A post-occupancy evaluation gathers information about the use of a building or environment from interviews and focus groups, might include carrying out an access audit after a building is completed or improvements have been made, and will analyse the data collected to give a picture of what works and what does not. The evaluation can give very useful information that can be used in the building studied and to inform future projects.

Even where a full evaluation is not carried out it can be useful to review the access audit recommendations after improvements have been made. There may have been other work carried out at the same time as the improvement work that has created more access problems; things may not have been implemented as originally intended and so might not be effective.

Information and training

Building manual

Where there is a manual covering issues related to the day-to-day running of a building, it should include information on the steps taken to achieve and maintain accessibility. If there is not a general building manual, an access manual could be set up.

The manual should include a copy of any access audit together with a record of specific measures taken to achieve or improve access. There should also be a permanent record of the specification of internal finishes, together with some explanatory notes, to ensure that issues of colour and luminance contrast are not forgotten when the building is redecorated. Information on maintenance of lifts or induction loops and guidance on good housekeeping, such as cleaning methods to ensure that floors are not made slippery or guidance on keeping corridors free of obstructions, should also be included.

Issues of accessibility should also be included in other relevant sections of any manual, such as those covering health and safety, security and emergency evacuation.

A manual will help to ensure that what has been learnt is not lost and that access becomes an ongoing concern.

Access guide

In some buildings it may not be possible to remove a physical barrier or it may be necessary, because of the nature of the business, to adopt practices and policies that do not allow independent access to a service or employment opportunity. For example, listed building consent may be refused for the installation of permanent ramps or a platform lift, requiring the use of temporary ramps on request; security arrangements appropriate to the service being offered may not allow unaccompanied travel around a building.

If there are unavoidable barriers to physical access which it would not be reasonable to remove, when judged on the merits of each individual case, making that information available in publicity material can allow a disabled person to plan his or her journey or visit and, where possible, enable arrangements to be made to remove or reduce the impact of such barriers.

An access guide, available in alternative formats, can advise all potential users of a service of any physical or managerial issues that might affect their access to the service. The guide can be sent out in advance to allow people to plan their visit, book a parking space or just be aware of areas where the access might be restricted.

Training

Staff training is critical to maintain access, to overcome shortcomings in building design and to provide accessible services and employment opportunities. Training can cover areas such as disability awareness and equality, use of equipment such as platform lifts and induction loops, British Sign Language, hearing awareness, clear lip speaking, guiding people with visual impairments and general access awareness.

Where access improvements are made to a building or accessible features incorporated in a new building, it is extremely important that staff understand how these improvements or features work in practice. If staff lack awareness, it is likely that full accessibility will not be achieved, whatever the design of the building.

> The interface between staff and customer is critical. It is here that the effectiveness of any accessibility practices and provisions will be tested, and where the disabled user will judge the reasonableness of any shortcomings in access to the service being offered.

Publicity

Where a good level of access has been achieved, either in a new building or following access improvements to an existing building, it can be useful to publicise this information. Contacting a local access group or disability organisation, or arranging publicity in a local paper, can encourage people to make use of accessible buildings and services. This, in turn, will encourage the maintenance of the accessible service and help ensure that the benefits of improving accessibility are visible.

3 Legislation, regulations and standards

Introduction

There are many responsibilities placed on designers, managers and owners of buildings by legislation and regulations relating to the accessibility of the built environment. The Building Regulations and the Disability Discrimination Act both have a major influence on accessibility and their requirements should be taken into account in the design and management of buildings and environments. Planning law and planning policy guidance cover issues that may affect access, as does other legislation in areas such as health and safety, occupier liability and fire precautions.

Legislation covering human rights and equal opportunities also influences the need for accessible environments. The Human Rights Act 1998 incorporates into UK law rights and freedoms guaranteed by the European Convention on Human Rights, and the Equal Treatment Directive covers access to employment and training.

BS 8300, which was published in 2001, amalgamates and revises earlier standards on access to buildings and gives detailed and thorough guidance on good practice. It is likely that this British Standard will be used as a guide to define 'reasonable provision' in relation to the Disability Discrimination Act.

This chapter looks at the relevant legislation, regulations and standards to identify the main issues affecting accessibility, disability and inclusion and to give sources of further reading and reference if required.

The Disability Discrimination Act 1995

The Disability Discrimination Act 1995 (DDA) introduced new laws and measures to combat discrimination against disabled people. The Act gives disabled people rights in the areas of:

- recruitment and employment;
- access to goods, facilities and services;
- buying or renting land or property.

In addition, the Act allowed the government to set access standards for public transport and required schools, colleges and universities to provide information on access to education for disabled pupils and students. The Special Educational Needs and Disability Act 2001 (SENDA) amended the DDA and expanded the duties relating to disabled pupils and students. These duties are explained in more detail later in the chapter.

The DDA is divided into a number of parts, defining disability and covering duties on different people or bodies:

- Part I – definition of disability;
- Part II – employers;
- Part III – providers of services, goods, facilities and those selling, letting or managing premises;
- Part IV – publicly funded education providers;
- Part V – public transport providers.

The duties imposed by each part are similar, but not identical. Each part is considered separately in this chapter.

The Act is supplemented by:

- regulations (secondary legislation);
- statutory Codes of Practice;
- guidance notes on the meaning of disability.

The DDA applies to the whole of the United Kingdom, including (as modified by Schedule 8) Northern Ireland. The Act does not apply to the Channel Islands or the Isle of Man and does not cover services or employment outside the UK.

Codes of Practice

Each part of the DDA is supported by a Code of Practice, which explains the principles of the law, illustrates how the Act might operate in certain situations and provides general guidance on good practice. The Code of Practice does not impose legal obligations, nor is it an authoritative statement of the law; however, it can be used in evidence in legal proceedings under the Act and courts and tribunals must take account of any part of a Code of Practice that is relevant to those proceedings.

The Codes of Practice give useful explanations of considerable parts of the law, but there are some areas of ambiguity and, ultimately, it is up to the courts to decide what is lawful.

Currently it is impossible to provide definitive answers to some important questions that might be asked about parts of the law in real life situations. The law is incomplete in some places, as some sections of the Act are not yet fully in place, and unclear in others. For example, the law does not clearly explain the basis for reasonableness and, currently, does not make clear the duties or obligations imposed on landlords to make the common parts of a building accessible for its tenants and its tenants' visitors.

Timetable for implementation

The rights granted to employees and job applicants under the DDA all came into force in December 1996. The part of the Act covering goods, facilities and services has been introduced in three stages:

- since December 1996 it has been unlawful to treat disabled people less favourably than other people for a reason related to their disability;
- since 1 October 1999 service providers have had to make reasonable adjustments for disabled people, such as providing extra help, making changes to the way a service is provided and providing auxiliary aids;
- from October 2004 it is intended that service providers will also have to make reasonable adjustments to the physical features of their premises to overcome physical barriers to access.

The implementation of duties affecting transport and education is covered later in the chapter.

The Disability Rights Commission

The government established the Disability Rights Commission (DRC) to help eliminate discrimination against disabled people and promote equality of opportunity. The DRC also advises the government on the working of disability legislation, such as the DDA, and sponsors test cases with a view to establishing case law. The DRC writes and produces the Codes of Practice relating to the DDA.

Relationship to building design

The DDA does not directly require buildings to be accessible to all disabled people and does not include standards for accessible building design; it is the services on offer within buildings that are the concern of the Act. Building designers should anticipate the needs of all building users, some of whom will fit the definition of disabled under the Act, and design accordingly.

> Even if a building is designed to be accessible, this will not in itself ensure that no discrimination is taking place. Conversely, a building may be badly designed and inaccessible to people with certain physical disabilities, but the way in which the services are provided may not be seen as discriminatory under the Act.

From 2004 the Act can require 'reasonable adjustments' to be made to physical features of buildings to overcome physical barriers to access. Although there is no requirement to make physical alterations before October 2004, it may be advisable to anticipate this requirement, particularly where other building works are planned, and remove barriers in advance of the duty coming into force.

However, there are a great variety of ways in which employers and the providers of goods, facilities and services can comply with the requirements if the Act and only some will involve alterations to premises.

The meaning of discrimination

It is necessary to understand what is meant by the term discrimination as defined by the Act. In essence, discrimination occurs where a disabled person is treated less favourably than a non-disabled person for a reason relating to their disability and without justification.

> Employment opportunities and services should be provided on an equal basis, this is not necessarily the same as an identical basis. Where there are constraints due to availability of resources or the use of existing buildings, identical opportunity may be a target but equal opportunity must always be the reality.

Actions that may be taken under the Act

A disabled person who believes that discrimination has taken place may bring civil proceedings in either an employment tribunal, for

Part II claims, or a county court for Part III. The DDA does not create any criminal offences. There are strict time limits for bringing actions. The time limit for a claim in an employment tribunal is 3 months and in the county court it is 6 months.

A disabled person who succeeds in an action may be awarded compensation for financial loss and injury to feelings. The court or tribunal also has the power to issue an injunction to prevent the discrimination re-occurring, and has other powers that can be used in employment disputes.

> There is no limit on the size of compensation awards that can be made. There are examples of awards in excess of £20,000 being made in employment cases.

Many cases of discrimination may be unintentional and the resulting dispute may be capable of being resolved by negotiation. The Disability Rights Commission sponsors an independent conciliation service to enable disputes to be settled without the need for recourse to the courts. If this is used the time limits described above may be extended.

DDA Part I: definition of disability

Part I of the Act defines what is a disability under the Act and, therefore, who is protected under it.

> It is intended that the Act should protect people who would generally be regarded as disabled. This is meant to be a fairly wide definition and extends well beyond the stereotype of a disabled person being someone using a wheelchair or a guide dog. To give some idea of numbers, the Labour Force Survey of 1998 shows that around 5 million people of working age judge themselves to be covered by the DDA definition.

The Act defines disability as 'a physical or mental impairment which has a substantial and long-term adverse effect on a person's ability to carry out normal day-to-day activities'.

- The term 'impairment' covers physical and mental impairments; this includes sensory impairments, such as those affecting sight or hearing, and cognitive impairment, such as learning disability.
- A long-term adverse effect is one which has lasted, or is likely to last, at least 12 months or for the rest of the person's life.
- Day-to-day activities are normal activities carried out by most people on a regular basis, such as washing, eating, catching a bus or turning on a television. The person must be affected in at least one of the respects listed in the Act: mobility; manual dexterity; physical co-ordination; continence; ability to lift, carry or otherwise move everyday objects; speech, hearing or eyesight; memory or ability to concentrate, learn or understand; or perception of risk of physical danger.
- The Act also applies to some people who have had a disability in the past, for example, someone who was disabled by a mental illness but has now fully recovered.
- Any treatment or equipment that alleviates or removes the effect of an impairment is ignored when considering whether a person has a disability. The only exception to this rule is where poor eyesight is improved by wearing glasses or contact lenses. In this case the effects that count are those that remain even with the glasses or lenses.
- People with severe disfigurements are covered by the Act, without any need to demonstrate that the impairment has a substantial adverse effect on their ability to carry out normal day-to-day activities.
- Certain conditions are specifically stated not to be impairments, including addiction to or dependency on alcohol, nicotine or any other substance, hay fever and self-imposed disfigurements.

DDA Part II: employment provisions

Under the DDA, it is unlawful for certain employers to discriminate against disabled people in their employment or when they are applying for a job. This includes arrangements and procedures such as application forms, interview arrangements, terms of employment, promotion or training opportunities, benefits and dismissal or redundancy. There is currently no duty on an employer with fewer than 15 employees. However, the government has announced that it intends to end this exemption from October 2004.

In addition, employers have a duty to make 'reasonable adjustments'. This applies where any physical feature of their premises, or any arrangements made by or on behalf of the employer, cause a substantial disadvantage to a disabled employee or job applicant. An employer has to take 'such steps as it is reasonable for him to have to take in all the circumstances' to prevent that disadvantage. Unjustified less favourable treatment and failure to make a reasonable adjustment which cannot be justified are seen by the Act as discrimination.

The employer's duty to make a reasonable adjustment comes into force when a disabled person applies for a job or an existing employee becomes disabled and requires an adjustment to be made. The duty is to an individual disabled person; it is not a general duty to disabled people at large.

Prison officers, fire fighters, members of the different types of police forces, employees who work wholly or largely outside Great Britain, members of the Armed Forces and employees who work on board ships, aircraft or hovercraft are not covered by the law currently. However, the government has announced that it intends to extend the employment provisions to cover all employers and occupations, with the sole exception of the Armed Forces, from 1 October 2004.

Adjustments that an employer might have to make

Examples given in the Act of adjustments that an employer might have to make are:

- making adjustments to premises;
- reallocating part of a job to another employee;
- transferring the disabled person to fill an existing vacancy;
- altering the person's working hours;
- assigning the person to a different place of work;
- allowing absences during working hours for rehabilitation, assessment or treatment;
- supplying additional training;
- acquiring special equipment or modifying existing equipment;
- modifying instructions or reference manuals;
- modifying procedures for testing or assessment;

When staff undergo appraisals or reviews as part of their employment it is important that everyone is given an equal opportunity to perform to the best of his or her ability, especially if promotion or pay increases depend upon the outcome.

If the appraisal process, or perhaps the room in which the interview takes place, does not allow someone who has a disability to perform as well as non-disabled people, it is likely that he or she is being discriminated against.

Facilities such as inductions loops, appropriate seating, adequate lighting, good acoustics and, if necessary, interpreters should be provided where needed to ensure everyone has the opportunity to maximise his or her performance in such important assessments.

After all – isn't that usually the purpose of the exercise?

- providing a reader or interpreter;
- providing additional supervision.

The Act lists a number of factors which may have a bearing on whether it will be reasonable for the employer to have to make a particular adjustment. The factors are:

- the effectiveness of the particular adjustment in preventing the disadvantage;
- the practicability of the adjustment;
- the financial and other costs of the adjustment and the extent of any disruption caused;
- the extent of the employer's financial or other resources;
- the availability to the employer of financial or other assistance to help make an adjustment.

> In practice, most adjustments will not involve alterations to physical features, often involve little or no cost or disruption and are therefore very likely to be considered reasonable.

Code of Practice   The Code of Practice for the Elimination of Discrimination in the field of Employment against Disabled Persons or Persons who have had a Disability gives guidance to employers on the implications of Part II of the Act.

> There is no minimum standard of adjustments to premises required by the DDA, but the employment Code of Practice gives examples of possible adjustments that it might be reasonable for an employer to make. These include: widening a doorway, providing a ramp or moving furniture for a wheelchair user; relocating light switches, door handles or shelves for someone who has difficulty in reaching; and providing appropriate contrast in décor.

Trade organisations   Trade organisations have similar duties to their members and applicants for membership as employers have under Part II of the Act. For this purpose, 'trade organisation' means an organisation of workers, an organisation of employers or any other organisation whose members carry on a particular profession or trade for the purposes of which the organisation exists.

If a trade organisation offers members services such as conference or training opportunities, they must be offered in a way that is accessible to all members who wish to attend. This would include the venue where the event is held, any accompanying notes, presentations and all other parts of the service.

DDA Part III: access to goods, facilities and services

Part III of the DDA deals with access to goods, facilities, services and premises. It is based on the principle that disabled people should not be treated less favourably simply because of their disability, by those who provide goods, facilities and services to the public.

Throughout this section reference is made to 'service providers' for convenience. Subject to certain exceptions, Part III of the DDA applies to any person or any organisation or entity that is concerned with the provision of services, including the provision of goods and facilities, to the public or a section of the public. It is irrelevant whether a service is provided on payment or without payment.

Services covered   The Code of Practice relating to Part III of the Act gives a list of some of the services that are covered. These include local councils, government departments and agencies, the emergency services, charities, voluntary organisations, hotels, restaurants, pubs, post offices, banks, building societies, solicitors, accountants, telecommunications and broadcasting organisations, public utilities, national parks, sports stadia, leisure centres, advice agencies, theatres, cinemas, hairdressers, shops, market stalls, petrol stations, telesales businesses, places of worship, courts, hospitals and clinics.

Some public bodies will be providing a service that may be covered by the Act in certain situations but not in others. For example, the police may be providing a service under the Act when giving advice about crime prevention, but are unlikely to be providing a service when arresting someone.

The Act says that services include 'access to and use of any place which members of the public are permitted to enter'. Thus a person who permits members of the public to enter such a place is providing a service to those people consisting of access to and use of that place.

Part III of the Act does not apply currently to the use of any means of transport, for example, taxis, buses, coaches, trains, aircraft and ships. However, transport providers are not wholly exempt from Part III. They still have duties to avoid discrimination and make reasonable adjustments to, for example, timetables, booking facilities, waiting rooms and other facilities at ferry terminals or bus, coach or rail stations.

When looking at transport, what needs to be determined is whether Part III or Part V of the DDA applies to the particular facility being considered. For example, Part V applies to vehicles whilst other obligations placed upon transport providers, as service providers, are generally covered by Part III.

A rule-of-thumb test is that if it moves (vehicles) it is usually covered by Part V and if it doesn't (infrastructure, information, services associated with transport) it is usually Part III that applies.

A wheelchair user is not protected under Part III of the Act if the bus on which she wishes to travel is inaccessible, but a service provided in the bus station, such as a café, is likely to be covered by the Act.

Further information is given on transport, and the proposed extension of Part III to include some transport services, in the section of this chapter covering Part V of the Act.

Education was excluded from Part III of the Act up until September 2002, when the Special Educational Needs and Disability Act 2001 removed the exemption. Further information is given on education in the section of this chapter looking at Part IV of the Act.

Services not available to the public   Part III of the Act currently does not cover services not available to the public, such as those provided by private clubs. However, where a club does provide services to the public, these will be covered.

A private golf club may refuse membership to a disabled golfer, but if the club allows non-members to use the course, a refusal to allow a disabled golfer to play is likely to be subject to the Act.

Exemptions that relate to private clubs only relate to clubs that are truly, and totally, private. Simply asking someone to join by signing in or completing a membership form as a prerequisite to joining in the activities taking place will not allow the club to claim this exemption. The Government is seeking to remove this exemption in the near future.

Part III of the Act does not cover manufacturer and design of products, as they do not involve the provision of services direct to the public.

Part III duties   The duties under this part of the Act are being introduced in three stages. The first rights of access came into effect from December 1996. Since then there has been a duty on service providers not to discriminate against disabled people by:

- refusing to provide, or deliberately not providing, a service which is offered to other people;
- offering a lower standard or worse manner of service; or
- offering less favourable terms.

From October 1999, service providers have been required to make reasonable adjustments to allow disabled people to use a service. This duty applies when access to a service is impossible or unreasonably difficult. The duty is being introduced in two stages. The first stage, which took effect from October 1999, requires service providers to take reasonable steps to:

- make reasonable adjustments to policies, procedures or practices which exclude disabled people or make it unreasonably difficult for disabled people to use the service, an example would be exempting working dogs from a 'no dogs' policy in a restaurant;
- provide auxiliary aids or services, such as the provision of information on audio cassette, which would enable disabled people to use a service; and
- where a physical feature is a barrier to service, finding a reasonable alternative method of delivering the service.

> The duties that came into force from October 1999 do not require service providers to do anything that would necessitate making a permanent alteration to the physical fabric of their premises. The duties affect the way in which buildings are used and managed. Duties relating to alterations to physical features come into force in 2004.

From October 2004, service providers will have a duty to take such steps 'as are reasonable in all the circumstances of the case' to modify physical features of premises which make it impossible or unreasonably difficult for disabled people to use the service. There are four different options which are available to a service provider to comply with this duty. In essence these are to:

- remove the feature;
- alter the feature so it no longer has the effect of making it impossible or unreasonably difficult for disabled people to use the service;
- provide a reasonable means of avoiding the feature; or
- provide a reasonable alternative method of making the service available to disabled people.

'Reasonableness' in service provision   Section 21 of the Act refers to a service provider being under a duty to take such steps as it is reasonable, in all the circumstances of the case, for it to have to take in order to make reasonable adjustments. What is a reasonable step will vary according to:

- the type of services being provided;
- the nature of the service provider and its size and resources;
- the effect of the disability on the individual disabled person.

It is not possible to assess reasonableness just in terms of cost, as this is only one issue to be taken into account. The Code of Practice relating to Part III of the Act gives a list of factors, which might be taken into account when considering what is reasonable:

- whether taking any particular steps would be effective in overcoming the difficulty that disabled people face in accessing the service in question;
- the extent to which it is practicable for the service provider to take the steps;
- the financial and other costs of making the adjustment;
- the extent of any disruption which taking the steps would cause;
- the extent of the service provider's financial and other resources;
- the amount of any resources already spent on making adjustments;
- the availability of financial or other assistance.

There is no exemption for small service providers, though it is likely that what is seen as a reasonable step for a large retailer with multiple outlets to take will not necessarily be seen as reasonable for a corner shop.

It is important to realise that there is no 'one size fits all' solution to meeting obligations under the DDA, for service providers or employers. Even an identical barrier to access in two identical buildings may well need to addressed in different ways, according to the several issues described above.

What is necessary is to consider each situation on its merits and not try to simply apply general solutions.

The Code of Practice contains examples of how the duties on service providers are intended to operate, and how the concept of reasonableness is to be interpreted in relation to the duty to make adjustments and provide auxiliary aids. It is clear that the duty is a continuing one, and is owed disabled people at large. It is described as 'an evolving duty, not something that simply needs to be considered once and once only, and then forgotten. What was originally a reasonable step to take might no longer be sufficient and the provision of further or different adjustments might then have to be considered'.

Service providers should plan ahead to anticipate the requirements of disabled people and the adjustments that may have to be made for them. They should not wait until a disabled person wants to use the service.

The Code of Practice also notes that service providers should bear in mind that there are no hard and fast solutions. 'Actions which may result in reasonable access to services being achieved for some disabled people may not necessarily do so for others. Equally, it is not enough for service providers to make some changes if they still leave their services impossible or unreasonably difficult for disabled people to use.'

The DDA expressly states that a service provider is not required to take any step 'which would fundamentally alter the nature of the service in question or his trade, profession or business'. Therefore, if a factor is inherent to the nature of the service being provided, such as low lighting in a nightclub, the service provider is not required to alter the service to make it fully accessible to disabled people. However, a decision of this sort would have to be defended if challenged.

Auxiliary aids and services   An auxiliary aid or service might be the provision of a special piece of equipment or simply extra assistance to disabled people from staff. The Code of Practice gives examples of auxiliary aids and services including the provision of information on audiotape and the provision of a sign language interpreter.

Reasonable steps in relation to auxiliary aids and services

The Code of Practice gives guidance on the range of auxiliary aids and services which it might be reasonable to provide to ensure that services are accessible to people with hearing disabilities:

- written information, such as a leaflet or a guide;
- a facility for taking and exchanging written notes;
- a verbatim speech-to-text transcription service;
- non-permanent induction loop systems;
- subtitles;
- videos with sign language interpretation;
- information displayed on a computer screen;
- accessible websites;
- textphones, telephone amplifiers and inductive couplers;
- teletext displays;
- audio-visual telephones;
- audio-visual fire alarms (not involving physical alterations to premises);
- qualified sign language interpreters or lipspeakers.

For people with visual impairments, the range of auxiliary aids and services which it might be reasonable to provide include the following:

- readers;
- documents in large or clear print, Moon or Braille;
- information on computer diskette;
- information on audio tape;
- telephone services to supplement other information;
- spoken announcements or verbal communication;
- accessible websites;

- assistance with guiding;
- audio description services;
- large print or tactile maps, plans and three dimensional models;
- touch facilities.

Auxiliary aids and services are not limited to aids to communication. A portable temporary ramp will also fall into this category, as will other items that assist physical access but do not involve a permanent alteration to the physical features or fabric of the building.

Physical features   A physical feature includes:

- any feature arising from the design or construction of a building on the premises occupied by the service provider;
- any feature on those premises or any approach to, exit from or access to such a building;
- any fixtures fittings, furnishings, furniture, equipment or materials in or on such premises.

All such features are covered whether temporary or permanent.

2004 duties   It is intended that, from 2004, where there is a physical feature that makes it impossible or unreasonably difficult for a disabled person to make use of a service, service providers will have to take reasonable steps to remove, alter or avoid it (for example, by installing a permanent ramp to enable wheelchair users to gain access to premises previously reached only by steps) if the service cannot be provided by a reasonable alternative method. The Act does not give any guidance on which of these approaches the service provider should take.

The Code of Practice, however, does state that service providers should adopt an 'inclusive' approach and consider first whether a physical feature which creates a barrier for disabled people can be removed or altered. The reasons given for recommending this approach are:

- the service will be available to everyone in the same way;
- it is preferable to any alternative arrangements from the standpoint of the dignity of disabled people;

- it is likely to be in the long-term interests of the service provider, since it will avoid the ongoing costs of providing the service by alternative means and may expand the customer base.

The Code of Practice recommends that only when removal or alteration is not reasonable should the service provider consider providing a means of avoiding the feature. If that is also not reasonable, the service provider should then consider providing a reasonable alternative method of making the service available to disabled people.

The law does not require a service provider to adopt one way of meeting its obligations rather than another. The Act is concerned with the end result and that the service is accessible to disabled people, rather than how this is achieved. However, if a service remains inaccessible, a service provider may have to defend its decision to adopt a certain approach.

If a service provider decides to use the option of providing a service by an alternative method and disabled people are then able to access that service without unreasonable difficulty, the obligations of the service provider under the Act will be satisfied. However, if it is still unreasonably difficult for the disabled person to use the service, the service provider would then have to show that it could not have reasonably removed or altered the physical feature, or provided a reasonable means of avoiding it. The cost of removing, altering or avoiding might be a relevant consideration. If a service provider takes no action it will have to show that there were no steps that it could reasonably have taken.

There are circumstances in which the duty to make reasonable adjustments is affected by compliance with the Building Regulations. More detail is given on this subject later in the chapter in the section on the Building Regulations.

Planning ahead   The Code of Practice suggests that it makes sense for service providers to plan ahead by taking any opportun-

Figure 3.1 If showers and other facilities are provided, they must be suitable for use by both disabled and non-disabled people – otherwise the provider (employer or service) will be providing the facility on a discriminatory basis.

ities that might arise, or bringing forward plans, to make alterations to their premises to benefit disabled people before 2004. It advises that although structural or other physical changes will not be required before 1 October 2004, they might be made before that date. The period leading up to October 2004 is intended to be a transitional period during which service providers can prepare for their new obligations.

> Service providers should use opportunities, such as a planned refurbishment programme prior to 2004, to remove or alter physical features that might create a barrier to access.

Reasonable adjustments in practice   Physical alterations will not always be the most appropriate way to improve the access to a service. Often measures such as disability awareness training for staff or simply allowing more time to serve disabled customers will be all that is required to make a service accessible. However, there will be situations when adjustments in the form of physical alterations are necessary.

The Code of Practice suggests that:

> 'regularly reviewing the way in which it provides its services to the public, for example via periodic disability audits, might help a service provider identify any less obvious or unintentional barriers to access for disabled people. Obtaining the views of disabled customers and disabled employees will also assist a service provider. Disabled people know best what hurdles they face in trying to use the services provided. They can identify difficulties in accessing services and might also suggest solutions involving the provision of reasonable adjustments.'

The role of access audits   The Code of Practice clearly suggests that service providers are more likely to be able to comply with their duty to make adjustments in relation to physical features if they arrange for an access audit of their premises to be conducted and draw up an access plan or strategy. It states that 'acting on the results of such an evaluation may reduce the likelihood of legal claims against the service provider'. However, undertaking an access audit is not a specific requirement of the DDA.

There may be situations where it is not reasonable for a service provider to anticipate a particular requirement. However, where a disabled person has pointed out the difficulty that he or she faces in accessing services, or has suggested a reasonable solution to that difficulty, it might then become reasonable for the service provider to take a particular step to meet these requirements.

A continuing and evolving duty   The duty to make reasonable adjustments is a continuing duty that should be regularly reviewed. The Code of Practice describes it as:

'an evolving duty, not something that simply needs to be considered once and once only, and then forgotten. What was originally a reasonable step to take might no longer be sufficient and the provision of further or different adjustments might then have to be considered.'

> Technological developments may provide new or improved solutions to certain problems and what was once an unreasonable step might become reasonable.
>
> Examples range from the use of personal items such as mobile phones to the use of Global Information Systems (GIS) by, for example, transport providers. The possibilities for technological developments are boundless – and costs are becoming more affordable.

Justification of less favourable treatment or failure to make reasonable adjustments In limited circumstances, the Act does permit a service provider to justify the less favourable treatment of a disabled person or a failure to make a reasonable adjustment. The justifications include health and safety considerations; incapacity to enter into a contract; the service provider being otherwise unable to provide the service to the public; enabling the service provider to provide the service to the disabled person or other members of the public and the greater cost of providing a tailor-made service.

> For a justification of discrimination to apply, the court must be satisfied that the service provider believed that one or more of the specified conditions existed and that it was reasonable for him or her to hold that opinion.

Health and safety The Act does not require a service provider to do anything that would endanger the health and safety of any person, including the disabled person in question. However, health and safety reasons that rely on prejudice or stereotyping of disabled people are no defence.

A cinema might not be justified in refusing entry to a wheelchair user based on an assumption that they would be a hazard in a fire. It is the responsibility of the management to make any special provision needed.

The Code of Practice explains that service providers should ensure that any action taken in relation to health and safety is proportionate to the risk. There must be a balance between protecting against the risk and restricting disabled people from using the service. Disabled people are entitled to take the same risks within the same limits as other people.

Before using health and safety to justify less favourable treatment, a service provider should consider whether a reasonable adjustment could be made to allow the disabled person to access the service safely.

Incapacity to contract   The Act does not require a service provider to contract with a disabled person who is incapable of entering into a legally enforceable contract or of giving an informed consent. However, the service provider should assume that a disabled person is able to enter into any contract and if there is a problem consider whether a reasonable adjustment could be made to solve it.

An example of a reasonable adjustment might be to prepare a contract document in plain English to aid understanding.

Service provider otherwise unable to provide the service to the public   A service provider can justify refusing to provide a service to a disabled person if this is necessary because the service provider would otherwise be unable to provide the service to other members of the public. However, this justification will only hold if other people would be effectively prevented from using the service. It is not enough that those other people would be simply inconvenienced or delayed.

An example given in the Code of Practice is if a tour guide refuses to allow a person with severe mobility impairment on a tour of old city walls because he or she has well-founded reasons to believe that the extra help the guide would have to give him or her would prevent the party from completing the tour. This is likely to be justified.

To enable the service provider to provide the service to the disabled person or other members of the public   A service provider can justify providing service of a lower standard or in a worse manner or on worse terms if this is necessary in order to be able to provide the service to the disabled person or other members of the public.

Again, before a service provider uses this condition as a justification it should consider whether a reasonable adjustment could be made to allow the disabled person access to the service.

Greater cost of providing a tailor-made service   If a service is individually tailored to the needs of a disabled person, the service provider can justify charging more for this service. However, justification on this ground cannot apply where the extra cost results from the provision of a reasonable adjustment.

Property owners and managers   The Act introduces particular duties on landlords and others who sell, let or manage premises. It is unlawful for a person with power to dispose of any premises to discriminate against a disabled person:

- in the terms of disposal of the premises;
- by refusing to dispose of the premises to the disabled person;
- in the treatment of the disabled person.

Disposing of the premises includes selling, letting, or assigning a tenancy. It does not include the hire of premises or booking of rooms in a hotel, as these would be covered by provisions relating to services.

In addition, those managing premises have a duty not to discriminate against a disabled person occupying the premises:

- in the way he/she permits the disabled person to make use of any benefits or facilities;
- by refusing to permit use of those facilities or by deliberately omitting permission;
- by evicting the disabled person or subjecting them to any other detriment.

It is also unlawful for a person whose licence or consent is required for the disposal of any leased or sub-let premises to discriminate against a disabled person by withholding that licence or consent.

As elsewhere in the Act, discrimination occurs when a disabled person is treated less favourably for a reason relating to his or her disability and when that treatment cannot be justified.

There is no duty to make reasonable adjustments to premises that are sold, let or managed. However, improving the accessibility of building stock may well increase potential value and ease of letting.

Exemptions in disposal of premises The provisions described above do not apply to owner-occupiers if:

- that person owns an estate or interest in the premises; and
- wholly occupies the premises.

However, if the owner-occupier uses the services of an estate agent or publishes an advertisement for the purpose of disposing of the premises, the Act applies.

Small dwellings exemption The provisions of the Act prohibiting discrimination against disabled people in the disposal of premises do not apply to certain 'small dwellings' (houses or other residential property). The Act gives a number of conditions that must be satisfied for the exemption to apply:

- the relevant occupier must live on the premises, intend to continue doing so and be sharing accommodation on the premises with other people who are not members of their household, for example, a multi-occupancy residential building with shared accommodation;
- the shared accommodation must not be storage accommodation or means of access;
- the premises must be 'small premises' as defined by the Act.

Justification of less favourable treatment in relation to premises   Less favourable treatment of a disabled person for a reason relating to disability amounts to discrimination unless that treatment can be shown to be justified. The Act sets out conditions that could apply to justify such treatment. These conditions are similar to the conditions that apply to justifying discrimination in the provision of services and the general approach to justification is the same.

Other issues covered by Part III of the DDA   The Act and the associated regulations affect the provision of particular services, such as insurance, guarantees and deposits. These issues are covered in the Code of Practice Rights of Access Goods, Facilities, Services and Premises.

Obtaining necessary consents

Nothing in the DDA removes the need for any necessary consent to be obtained where this is required under general law or under a contract or lease, but in some cases regulations set out what sort of action is reasonable in terms of seeking such consent. Regulations also cover landlords' duties to consent to a tenant's application to carry out adjustments in order to fulfil a duty under the DDA.

The regulations relating to Part II already apply; the Part III regulations do not come into force until October 2004. The various Codes of Practice contain useful explanations of how the regulations operate.

Consents may also be needed from mortgagees or neighbours, for example, an adjoining owner with the benefit of a restrictive covenant preventing alterations being made. In these cases the employer or service provider must seek consent but need not make any alteration until that consent has been obtained.

There are extensive provisions relating to the obtaining of land-lord's consent. The Codes of Practice set out the procedures in considerable detail, but in summary:

- whatever the lease states, it is to be read as if the tenant is permitted to make any adjustments required by the DDA with the consent of the landlord, and the landlord is not permitted to withhold consent unreasonably or impose unreasonable conditions;
- regulations set out when the landlord will be acting reasonably and the conditions that a landlord may reasonably impose.

Where a landlord unreasonably refuses consent, or imposes unreasonable conditions, the duty of the employer or service provider is to inform the disabled person of this. The matter then becomes a dispute between the disabled person and the landlord, and the landlord can be made party to the disabled person's application to the court or tribunal. The court or tribunal has the power to authorise the employer or service provider to make alterations, as well as ordering the landlord to pay compensation.

Statutory consents

A service provider may have to obtain statutory consents such as planning permission, building regulation approval, listed building consent or fire regulations approval, before carrying out altera-tions to physical features. The Act does not override the need to obtain such consents. Where consent is refused, there is likely to be a means of appeal. Whether the service provider has a duty to appeal will depend on the circumstances of the case.

Applying for consent is always a reasonable step to take, but the employer or service provider might have to consider whether it is reasonable to make a temporary adjustment in the meantime, or adopt a different permanent adjustment not requiring consent.

Building Regulations and the DDA

The duty under the DDA to make reasonable adjustments to physical features of buildings can be affected by building regulation compliance. Where the physical features of a building met the requirements of Part M of the Building Regulations (or its equivalent in Northern Ireland or Scotland) at the time of its construction, and continue to meet them, an employer or service provider may not have to make any further adjustment to those features. This might apply, for example, to the width of a doorway. However, the employer or service provider might still need to alter other aspects of the door, such as the handle.

> Even if the exemption applies so that no physical adjustment is required, there may still be an obligation to make an alternative adjustment if a reasonable one can be identified, or to provide the service in another manner.

This exemption differs in relation to the duties under Part II and Part III of the Act. In particular, the exemption under Part III is only available for a period of 10 years from the date when a feature was constructed.

> It would be unfortunate if designers opted for the simple solution of following the guidance in Approved Document M in every case, in order to be sure of gaining future exemption, rather than developing more creative solutions to access issues that may be more suitable for the particular building or circumstance.
>
> It must also be remembered that the exemption only applies to those elements of a building to which Part M applies. Therefore, items such as door handle design, floor finishes, colour contrast, lighting, surface finishes, acoustics etc, are not subject to the exemption.

DDA Part IV: education

Part IV of the DDA requires education institutions in England and Wales to inform parents, pupils and students about their arrangements for disabled people. From January 1997 the governing bodies of all maintained schools, except special schools, have had to publish in their annual reports:

- a description of the admission arrangements for disabled pupils;
- details of the steps taken to prevent disabled pupils from being treated less favourably;
- details of facilities provided to assist access to the school for disabled pupils.

The Act also places new duties on Further Education Funding Councils and Higher Education Funding Councils.

Part III of the DDA initially excluded publicly funded education from the provisions relating to goods, facilities and services. The exemption of education was effectively removed by the Special Educational Needs and Disability Act 2001, which came into force in September 2002.

Special Educational Needs and Disability Act 2001   The Special Educational Needs and Disability Act 2001 (SENDA) amended Part IV of the DDA and expanded the duties relating to disabled pupils and students. It also removed the exemption of publicly funded education from Part III of the Act, though where a Part IV duty applies, Part III cannot apply. Education providers are now required to make 'reasonable adjustments' for disabled students and pupils. These new duties cover all areas of education, schools, colleges, universities, adult education and youth services, and include:

- a duty not to treat disabled students/pupils less favourably than non-disabled students/pupils without justification;
- a duty to make reasonable adjustments to policies, procedures and practices that may discriminate against disabled students/pupils;

- a duty to provide education by a 'reasonable alternative means' where a physical feature places a disabled student/pupil at a substantial disadvantage (there is no general duty to remove or alter physical features or provide auxiliary aids or services);
- a duty on local education authorities in England and Wales to plan strategically and increase the overall accessibility to school premises and the curriculum.

Post-16 education providers have further duties, which come into force as follows:

- from September 2002 they have had a duty not to discriminate against existing and prospective disabled students by treating them less favourably than others in the provision of student services;
- from September 2003 they have had a duty to make reasonable adjustments and provide auxiliary aids;
- from September 2005 it is intended that there will be a duty to make adjustments to physical features.

The duty to make adjustments to physical features is an anticipatory and continuing duty. Further and higher education providers will be required to consider the needs of disabled people in general and not wait until an existing or prospective student requires a reasonable adjustment to be made.

DDA Part V: transport

Part V of the DDA deals with public transport vehicles. It allows the government to set accessibility standards for buses, coaches, trains, trams and taxis.

The regulations set minimum technical requirements to ensure that disabled people can use public transport safely and comfortably. Areas covered include size of door openings, dimensions of seating compartments, colour contrast, ramps on platforms to allow access to trains and accessible WC facilities on trains.

Proposed extension of the Part III provisions The government is proposing to extend the Part III provisions of the

DDA to include services such as rail vehicles, buses, coaches, taxis, private hire vehicles, aviation, shipping, car hire services and breakdown services. It proposes to:

- make it unlawful to discriminate against a disabled person in refusing to provide a service which is provided to other members of the public, or in providing a service of a lower standard or on less good terms than those available to other members of the public;
- require, where reasonable, changes to any 'practice, policy or procedure' which makes it impossible or unreasonably difficult for a disabled person to make use of the service;
- require where reasonable, the provision of an auxiliary aid or service which would enable a disabled person to make use of a service available to other members of the public.

The proposed changes would require primary legislation before they could be implemented and there would be a lead-in time before any changes were brought into force.

The Building Regulations

The Building Regulations provide functional requirements for building design and construction. They exist to ensure the health and safety of people in and around all types of buildings and also provide for energy conservation and access.

The regulations themselves are very short and contain no detail; this is given in a series of Approved Documents which provide guidance as to how the requirements of the Building Regulations may be met. It is not mandatory to comply with this guidance; other methods of construction or design may be equally acceptable provided the Building Regulation requirements are met. The Building Regulations are enforced by local authorities or approved inspectors.

The requirements of the Building Regulations that directly relate to access are Part M – Access and facilities for disabled people, and Part B – Fire safety.

Part M - Access and facilities for disabled people

Part M of the Building Regulations relates to access and facilities for disabled people. It applies to new buildings, some extensions, and features outside a building which are needed to provide access to the building from the edge of the site and from car parking within the site. The Approved Document states that the requirements of Part M will be met by making it reasonably safe and convenient for disabled people to:

- gain access to and within buildings other than dwellings and to use them. The Approved Document explains that these provisions are for the benefit of disabled people who are visitors to the building or who work in it.
- visit new dwellings and to use the principal storey. The provisions are expected to enable occupants to cope better with reducing mobility and to 'stay put' longer in their own homes, though not necessarily to facilitate fully independent living for all disabled people.

Approved Document M   Approved Document M gives guidance on part M of the Building Regulations. The document sets out a number of objectives to be met, covers design considerations and gives technical details of design solutions. These solutions, called provisions in the document, show one way in which the requirements might be met.

> The provisions given in Approved Document M do not show the only way of meeting the requirements, they merely illustrate one way and there is no obligation to adopt any of them. Alternative solutions may well be acceptable provided the overall objective is met.

Proposed changes   At the time of writing, Part M is under review and a new edition is due to be published in late autumn 2003 to take effect from spring 2004. The proposed changes fall into three main categories:

- updating to take account of the guidance in BS 8300:2001;
- bringing Part M into line with other Parts of the Building Regulations by extending its scope to include alterations to existing buildings and certain changes of use; and
- application of the concept of access and use for all.

It is proposed that specific references to disabled people will be omitted from the Requirements and references in the Approved Document will be broadened to include 'people, including parents with children, elderly people and people with disabilities'.

The guidance in Approved Document M is to be reordered and overhauled to reflect the recommendations in BS 8300:2001.

The proposals set out that, if Part M applies, reasonable provision should be made in buildings other than dwellings:

- for people, including disabled people, to be able to reach the principal entrance to the building, and certain other entrances, from the edge of the site and from on-site car parking;
- so that elements of the building do not constitute a hazard for a person with a sight impairment;
- for people, including disabled people, to have access into and within any storey of the building and to the building's facilities;
- for suitable accommodation for people in wheelchairs, or people with other disabilities, in audience or spectator seating;
- for aids to communication for people with a sight or hearing impairment in auditoria, meeting rooms, reception areas and ticket offices; and
- for sanitary accommodation for the users of the building, including disabled people.

The proposals also cover access to dwellings and contain information on access to educational establishments and historic buildings.

It is also proposed that an Access Statement should be provided for non-domestic buildings where a designer or developer wishes to depart from the guidance in the Approved Document. The statement should set out the reasons for departing from the guidance and the rationale for the design approach adopted. It is intended to give an opportunity to provide justification where full access is not achieved in works to existing buildings, particularly historic buildings, due to the constraints of the existing structure and to propose compensatory measures and to identify areas where access is restricted or not required.

Part B – Fire safety

Part B of the Building Regulations applies to all construction, including new-build, refurbishment, extensions and alterations and sets out the requirements for fire safety. Approved Document B gives guidance on meeting these requirements and makes reference to BS 5588, which gives detailed information on the design, construction and use of buildings.

In the current Approved Document B to the Building Regulations, only one paragraph offers guidance on appropriate means of escape for disabled people, which is as follows:

'It may not be necessary to incorporate special structural measures to aid means of escape for the disabled. Management arrangements to provide assisted escape may be all that is necessary.'

BS 5588 Fire precautions in the design, construction and use of buildings

BS 5588 covers different building types and elements of buildings. It is divided into a number of parts and BS 5588-8 covers means of escape for disabled people and introduces the concepts of refuges, which are designated safe areas where disabled people can wait in the event of a fire, and evacuation lifts. Further information on emergency evacuation is given in Chapter 4.

It is proposed that BS 9999, which is out for consultation at the time of writing, will replace BS 5588. It is intended that this new Code will offer the most practical, relevant and up-to-date guidance to assist designers and managers of buildings in providing and managing reasonable means of escape for all building users.

BS 9999 Code of practice for fire safety in the design, construction and use of buildings

BS 9999 addresses issues that need to be considered during the design process as well as those that apply when a building is in use and when alterations are made. It will cover fire safety management, means of escape, structural protection of escape facilities and the structural stability of the building in the event of a fire, and the provision of access and facilities for fire fighting. The guidance considers the nature of the occupants and the use of the building, as well as the likely fire growth and the resulting risks associated with that use, to establish a risk profile for the building. This will allow a more flexible approach to fire safety design to be developed taking account of varying physical and human factors.

BS 8300:2001 Design of buildings and their approaches to meet the needs of disabled people – Code of Practice

BS 8300:2001 is an amalgamation and updating of BS 5619:1978 and BS 5810:1979, and gives detailed guidance on good practice in the design of domestic and non-domestic buildings. Importantly, the guidance also draws on research, commissioned by the Department of the Environment, Transport and the Regions in 1997 and 2001, into the access needs of people with disabilities. The research looked into issues such as reach ranges and space requirements in order to assess the capabilities and needs of people in relation to the use of buildings. The guidance incorporates the research findings and gives detailed design recommendations set in context by a commentary explaining user needs.

BS 8300:2001 contains sections covering building elements as well as particular building types and the guidance given takes account of a wide range of needs. The British Standard gives recommendations on car parking, access routes to and around buildings, entrances and interiors, horizontal and vertical circulation, surfaces and communication aids, facilities in buildings, assembly areas, individual rooms and building types.

The proposed new edition of Part M of the Building Regulations incorporates some of the guidance and standards given in

BS 8300:2001. It is likely that the guidance given in the British Standard will be taken into account when considering 'reasonable provision' in relation to the Disability Discrimination Act.

Planning legislation and guidance

The current system of planning is based on legislation, orders and regulations. In general, the legislative framework is provided for by the following three pieces of legislation:

- The Town and Country Planning Act 1990
- Planning (Listed Buildings and Conservation Areas) Act 1990
- Planning and Compensation Act 1991.

Under legislation, every local planning authority is required to prepare a development plan for their area. At county level this may take the form of a structure plan, at district or borough levels a local plan or, for single unitary authorities and London boroughs, a unitary development plan.

There is a requirement for all local authorities to ensure that such plans are always updated and continue to be relevant to planning applications.

In the preparation of a development plan, a local authority must take into account any national and regional planning guidance that is relevant at the time. Such guidance is mainly identified in a series of Planning Policy Guidance notes (PPGs) and in Regional Planning Guidance notes (RPGs).

Those PPGs that currently relate to accessibility and inclusive environments include:

- PPG1: General Policies and Principles
- PPG3: Housing
- PPG6: Town Centres and Retail Developments
- PPG12: Development Plans
- PPG13: Transport
- PPG15: Planning and the Historic Environment
- PPG17: Planning for Open Space, Sport and Recreation
- PPG25: Flooding.

It must be remembered that the revisions to the content of PPGs is ongoing and it is important to keep abreast of any additions or alterations in the guidance offered.

In addition, local planning authorities have a duty under the Town and Country Planning Act to draw the attention of developers to the relevant BS covering access for disabled people.

The government has issued a good practice guide, Planning and Access for Disabled People (ODPM 2003), which addresses the need for the planning system to enhance and enforce inclusive design provisions by those applying for planning consent. The guide describes how all those involved in the planning process can play their part in delivering a physical environment that can be used by everyone and explains the respective roles of the planning system, Building Regulations and the DDA.

Occupier Liability Acts

Who is an occupier?

The occupier of a premises can be defined as a person who has a:

'sufficient degree of control over premises that he ought to realise that any failure on his part to use care may result in injury to a person coming lawfully there'. (Lord Denning)

The person who is 'coming lawfully there' is defined as a visitor.

Occupier Liability Act 1957 (OLA 57)

Under the OLA 57, the occupier owes a single common duty of care to all his or her visitors. This common duty of care is a duty to take such care as in all the circumstances of the case is reasonable to see that the visitor will be reasonably safe in using the premises when he or she is invited or permitted by the occupier to be there.

In general, the standard of care that can be expected of an occupier is the same as that in an ordinary action of negligence. Therefore, whilst the occupier would not be expected to guard against improbable or unlikely events, he or she must do sufficient to exercise the common duty of care described above. In addition, it is not the premises but the visitor who must be made reasonably safe, and the precautions needed to fulfil this duty will vary according to the particular visitor.

Importantly, where a landlord retains control of the access to premises or to the common parts of a property, he or she is treated as the occupier and thereby owes the common duty of care to the tenants, their families and their visitors.

Whilst an occupier may discharge his or her duty to a visitor by giving a warning of any danger that may exist, the warning may not be taken as absolving the occupier unless in 'all the circumstances' it was enough to enable the visitor to be reasonably safe. The simple presence of a warning does not necessarily suffice.

Occupier Liability Act 1984 (OLA 84)

Under the OLA 84, an occupier may owe a duty to persons other than his visitors, often referred to as non-visitors. This may include for example, trespassers, people who enter the land or the area occupied in the lawful pursuance of their job or people lawfully exercising a private right of way. The duty is to take such care as is reasonable in all the circumstances of the case to see that the 'non-visitor does not suffer injury on the premises because of a danger that may exist there'.

What constitutes reasonable care will vary with circumstances. These varying circumstances might include the way the building or space was entered, the age of the person entering, the type and nature of the premises, the extent of the risk and the cost of carrying out precautionary work.

Relationship of the Occupier Liability Acts to inclusive design and access management

Employers, service providers and others may well have responsibilities, relating to accessibility, under the Occupier Liability Acts, as well as under the DDA. For example, in existing buildings it is common to find an accessible WC that is not provided with an emergency alarm call. The DDA identifies that it should not be unreasonably difficult to use a service – but is the lack of an alarm in an accessible WC making that WC any more difficult to use than the general provision WCs? In most cases it could be argued that to specify the provision of an alarm could be seen as a reasonable provision, the limited cost and the inconvenience of doing so being unlikely to make such an action unreasonable for most service providers.

As occupiers, the service provider will owe a duty of care to visitors 'to take such care as in all the circumstances of the case is reasonable to see that the visitor will be reasonably safe in using the premises'.

With regard to the use of an accessible WC facility, it is known, for example, that disabled people can experience some difficulty, and even danger, when transferring between the wheelchair and the WC. If they fall and are unable to call for assistance, they may be in the toilet for some time before being discovered. What must be considered in each individual case is whether the occupier is taking such care as 'in all the circumstances of the case' is reasonable to ensure that visitors are safe when using the premises.

There are several examples within the built environment to which this could be applied. For example, it could be argued that slippery, shiny floor finishes do not necessarily discriminate against disabled people because all users experience difficulties when using them. However, disabled people may certainly find the use of such floors more difficult, restrictive and dangerous than other people. It could be argued that a service provider needs to take into account when considering 'all the circumstances of the case' that the needs of disabled people using his or her service, who might be more vulnerable to slipping and falling, are greater than those of other people and should be addressed accordingly.

The examples given above are by no means the only areas that may need to be considered. Each issue should be considered on its merits. The duty owed 'in all the circumstances of the case' to disabled people who may be non-visitors under the OLA 84 may also be different to that owed to others – especially if a difference in the level of ability could be foreseen.

In that respect, the recommendations made to address some issues identified in an access audit may well be based upon the responsibilities of the service provider or employer under legislation other than the DDA (e.g. OLA, Health and Safety) – and the audit report should identify the items that relate to either discrimination or minimum legal management standards.

An access audit, and the recommendations which arise from it, should clearly identify those issues which relate to discrimination under the DDA and those issues which relate to appropriate management practices by an occupier or employer under other legislation.

4 Design criteria

Introduction

Access to buildings and environments, and the services they house, is easier for everyone if certain basic criteria are met. These criteria cover issues that affect all building users, not just specific groups of people.

It is not necessary to separate the needs of a wheelchair user from those of an ambulant person – both need to be able to find the entrance to a building, enter through the doorway and approach the reception desk, both may need to use a lavatory or a car park. If people are divided into groups based on differing needs, it is apparent that the areas of overlap far exceed the specific requirements for each group. The design guidance given here is intended to cover the needs of as wide a range of users as possible.

This approach does not deny that there are specific areas where particular assistance can be provided. Hearing enhancement systems, such as induction loops, or the provision of information in Braille, are useful to certain building users. Specific provisions that meet particular needs should be part of inclusive design.

The guidance given here can be used when designing new buildings or taken as a standard to assess and improve existing buildings. However, it is important that it is seen in context – inclusive design is not simply the application of a set of criteria. The effective implementation of much of the guidance depends upon management understanding and input. To be effective, inclusive design should be seen as an approach that is an integral part of the design process and continues throughout the life of a building or environment. Issues of management, maintenance and building use will affect whether a good level of access, and inclusiveness, is achieved and is sustained in use.

The list of design criteria that follows is not exhaustive – it is intended to cover those areas that most often cause problems for users. Sources of greater detail, where appropriate, are noted in the text and listed in Appendix C.

External environment

Car parking

Car parking or setting down are often the first activities that take place on arrival at a public, commercial or employment building and can have a great effect on ease of access. Issues such as signs, travel distances from car park to entrance and design of external routes affect everyone, not just people with disabilities. Depending on the building use consideration should be given to providing priority parking spaces for anyone who needs to park nearer the entrance, for example parent and child parking, in addition to designated spaces for disabled people.

- Car parking should be provided close to the entrance to the building and, especially on sloping sites, designated parking provision for disabled people should be at the same level as the entrance.
- Designated car parking provision should be signposted from the entrance to the car park and the bays clearly identified.
- The number of designated parking spaces should be appropriate for the type and use of building.

Figure 4.1 Car park showing excellent use of signage, ground markings and pedestrian routes.

- Designated parking spaces should be within 50 m of the principal entrance to the building; where the route is covered this could be increased to 100 m.
- Bays should be of sufficient size to allow doors and boot lids to be fully opened and to allow drivers and passengers to transfer to a wheelchair. A single accessible bay should be a minimum of 4800 mm long by 3600 mm wide. If a bank of bays is provided, each bay should be at least 4800 mm long by 2400 mm wide with an additional 1200 mm transfer zone between bays. A 1200 mm wide safety zone is recommended for boot access and use of rear hoists.

Where there is insufficient space to make all disabled parking bays long enough to allow for rear exit and loading, a marked-out pedestrian route on the road behind the parking spaces can double up as loading space.

Figure 4.2 Parking bay with adequate room for both side and rear exit and entry to allow easy use. Although the rear exit and entry area is not marked, it can clearly be achieved without encroaching into the vehicle route.

- Pedestrian routes in car parks should be clearly defined.
- Where there are kerbs between parking areas, transfer areas and routes to the building, they should be dropped to allow easy passage for wheelchair users, with tactile paving provided at all road crossing points.
- The car park surface should be smooth and even.
- Designated parking for disabled people in multi-storey car parks should be at the same level as the principal accessible entrance to the building or the main access route to and from the car park. There should be signs indicating the accessible route to ticket machines, lifts and exits and, where appropriate, the building that is being visited.
- Where there are ticket machines, coin- or card-operated barriers or other controls, these should be at an appropriate height and design to allow easy use by everyone.
- It is important that management procedures are in place, and in use, to ensure that other motorists do not use the bays designated for disabled people to use.

(a)

(b)

(c)

Figure 4.3 (a–c) Good clear signage, correct drainage, and firm non-slip surfaces are essential. Routes between parking areas should be clear and well defined; bollards can prevent encroachment.

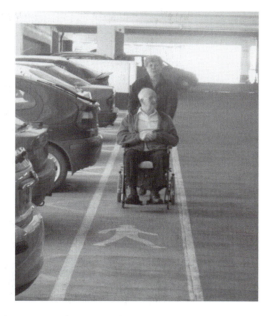

Figure 4.4 Parking is at the same level as the entrance to the shopping area with a clearly defined route.

Recommended numbers of designated spaces for disabled people are given in Inclusive Mobility (DfT 2002) as follows:

- for car parks associated with existing employment premises – 2% of the total capacity, with a minimum of one space (this figure does not include spaces for disabled employees);
- for car parks associated with new employment premises – 5% of the total parking capacity (to include both employees and visitors);
- for car parks associated with shopping areas, leisure or recreational facilities and places open to the general public – a minimum of one space for each employee who is a disabled motorist, plus 6% of the total capacity for visiting disabled motorists.

The number of designated spaces may need to be greater at buildings that specialise in accommodating groups of disabled people, such as some hotels and sports stadia.

Figure 4.5 Setting down and pick-up points should be clearly marked with dropped kerbs. Care should be taken to ensure bollards do not obstruct routes.

Setting down

A setting down point is useful for people arriving by car, taxi or dial-a-ride bus.

- A clearly sign-posted setting down point on a level surface close to the main entrance of the building should be provided, with a shelter if possible.

> It is essential to ensure that dropping off points are properly signed and managed so that they are not used for other 'quick activities' such as the delivery of goods, etc.

External routes

External routes around buildings should be designed to allow easy, unobstructed access for everyone. Routes should be clearly

Figure 4.6 Seating can be useful on longer routes. Armrests can help people when sitting or standing.

defined, wide enough for all users, free from hazards and have firm, even, slip-resistant surfaces.

Surfaces   Uneven surfaces, loose material such as gravel and wide joints between paving units can all cause problems for people using external routes. For wheelchair users, people with visual impairments and people who are unsteady on their feet, such surfaces can be hazardous or prevent use of the route.

- Surfaces should be firm, even and consistent and have a slip-resistant finish in all weather conditions.
- The material used to form the path should be capable of withstanding the loads and volume of traffic it will carry.
- Path surfaces containing grass, unless supported by a suitable reinforced mesh, gravel or deeply ridged concrete sets should be avoided.
- Joints between paving units should not exceed 10 mm in width and 5 mm in depth. Any difference in level between adjacent paving units should be no greater than 5 mm.
- The edges of paths should be provided with some method of assisting blind or partially sighted users, for example, a

Figure 4.7 Good surfaces and level entrances are necessary, but management control of potential obstructions is also required.

raised kerb or tapping rail, though care should be taken to ensure that this does not represent a tripping hazard for other users.

- Pedestrian and vehicular routes should be clearly distinguished using texture, colour or a small change in level, with dropped kerbs where appropriate.
- Changes in texture and profile of paving can be used to give information to pedestrians. A surface that is uncomfortable to walk on can be used to steer people away from a hazard or a route across a large open space can be identified by a difference in surface.

An accessible route can be formed by laying a smooth path of stone or concrete paving across an existing uneven surface, such as cobbles or gravel.

Figure 4.8 A smooth path can be laid across an uneven surface to provide an accessible route.

Tactile paving   There are a number of specific profiles of paving that have been developed to give particular information to pedestrians.

- The modified blister is used externally at a dropped kerb or raised road surface to indicate the edge of the footway and the start of the carriageway.
- Corduroy paving, which has a half-rod-shaped profile in a ribbed pattern, is used to provide a hazard warning. Part M of the Building Regulations requires this to be used at the top and bottom of external flights of steps and at intermediate landings where there is access onto the landing other than from the steps (see *External steps and stairs*).
- A flat-topped ribbed profile is used to denote a guidance path. This could be used in a large open space such as a pedestrian precinct.

There are also specific profiles for use at platform edges and on segregated paths for cycles and pedestrians. Detailed information

Figure 4.9 Good use of a tactile paving in terms of colour (buff for an uncontrolled crossing) and profile (blister).

on tactile paving is given in Guidance on the Use of Tactile Paving Surfaces (DETR 1998).

> Where tactile paving is used it should be of the correct profile, colour and hardness to give the appropriate message. For people who rely on tactile clues to gain information about the environment, being given the wrong information is often much worse than being given no information at all. If there is no information, they will be alert to potential problems; incorrect information could give a false sense of security and be potentially dangerous.

Path width and gradient   The width of paths should be adequate to meet the needs of all users. People using wheelchairs, pushing buggies or using walking aids should be able to pass each other easily. On narrower paths passing places should be provided.

Figure 4.10 A good approach to a building can greatly enhance people's perception of it – as well as its accessibility.

- The recommended minimum width for paths is 1200 mm, though 1500 mm is preferred. A width of 1800 mm will allow wheelchairs and pushchairs to pass each other.
- Path gradient should be less than 1 in 20. A path with a gradient of 1 in 20 or greater should be treated as a ramp with appropriate handrails and landings.
- The cross fall on a path is also important, especially for people using wheelchairs, and should not exceed 1 in 50.

A dropped kerb can allow a blind or partially sighted person to leave the pavement unknowingly; tactile paving should always be provided to give information about the lack of a kerb and the potential danger ahead.

Figure 4.11 A barrier to mobility?

Changes of direction   Care should be taken at junctions and corners of paths to ensure that any change of direction can be undertaken easily, safely and with minimum effort from the user. Ambulant disabled people, wheelchair and guide dog users and those pushing prams or buggies will have certain width requirements in negotiating changes of direction and these must be allowed for in any path design.

- Wherever possible corners at changes in direction should be splayed or rounded.

Gratings   Drainage gratings can cause tripping, slipping and trap wheelchair wheels, ends of sticks, crutches and the heels of shoes. Where feasible they should be positioned off routes. If they are located within paths, they should be flush with the surrounding surface.

- Slots in grating should be less that 13 mm wide and set at right angles to the direction of travel.

Figure 4.12 Wide slots in drainage gratings can catch wheels, sticks and heels.

- Circular holes should not exceed 18 mm diameter.
- Dished channels can cause tripping and should not be provided on access routes. Where a drainage channel is incorporated at a dropped kerb, there should be a flat plate across the channel for the length of the dropped section.

Dropped kerbs   At pedestrian crossing points and other places where level access is required between path and a carriageway, for example in a car park, there should be a dropped kerb.

- The width of the dropped section should be at least 1200 mm, though 2000 mm is preferred.
- Gradients of paving at dropped kerbs should not exceed 1 in 15.
- It is very important to provide the correct profile, colour and layout of tactile paving to give the appropriate information at

Figure 4.13 Potential tripping hazards should be highlighted using colour and luminance contrast or, wherever possible, removed altogether.

road crossings. See Guidance on the Use of Tactile Paving Surfaces (DETR 1998).

Street furniture   Street furniture such as litterbins, signposts, bollards and seating should be carefully designed and placed so as not to obstruct routes or restrict widths and should be clearly visible. Items could be grouped and clear routes identified by textural changes in paving.

- Paths should be kept clear of obstructions at ground level, such as litterbins or planters, or any projections into the walking zone such as projecting signs or overhanging tree branches.
- Bollards should be a minimum of 1000 mm high and adequately colour contrasted with the background against which they will be viewed.

Figure 4.14 Projections into the walking zone can be very dangerous for all people, but especially people with visual impairments. Projections should be avoided, or at least properly protected, at all times.

Figure 4.15 Seating should be located where it does not obstruct routes and should be clearly visible.

The visibility of potential obstacles is important to many people. People who are deaf or hard of hearing and who are talking and looking at a companion, or communicating using lip reading or signing, will be relying on their peripheral vision to identify obstacles. Appropriate colour and luminance contrast of potential obstacles is essential for them, just as it is for a person with a visual impairment or anyone who is simply walking along not paying attention to their surroundings.

Handrails   Handrails should be provided at changes of level, or as guardings, and can be used to guide people away from obstructions in the walking zone.

Lighting   Wherever possible, all external routes should be well illuminated, without strong shadows and dark areas.

- Changes of level and other potential hazards should be well lit.
- Lighting should not give glare or cross shadows, particularly around potential hazards such as steps or ramps.
- Light sources should not reduce colour definition, which has been introduced to provide information or identify hazards.

To be effective, external lighting needs regular maintenance. Fittings should be kept clean to maximise available light, and bulbs and fittings swiftly replaced when damaged or broken.

External ramps

The provision of appropriately designed, constructed and managed ramps is of importance to all users, but especially those using wheelchairs, pushing buggies or trolleys, and people using walking frames.

> Ramps should only be considered where it is necessary to address unavoidable changes in level and never used simply to overcome changes of level that could have been avoided by good, thoughtful design.

Ramps that are long and steep may cause difficulties, especially for people using wheelchairs. A wheelchair user or companion pushing the wheelchair might not have sufficient strength to travel up a steep gradient or, importantly, to control the journey down. Steep gradients, especially short steep gradients such as those found at doorways and kerbs, can also cause wheelchairs to tip up backwards when going up and increase the danger of a wheelchair user falling forwards when coming down.

Landings are needed as resting places, but travel along long ramps can become tiring even with landings. It is the overall length of travel on the ramp that is important, not the number of places available to allow someone to rest. Ramps should be designed to appropriate gradients, but with the overall length kept to an absolute minimum.

Ramps that cut across stairs, resulting in tapered risers, can be hazardous for people using the stairs and the ramp and should be avoided.

> Some people prefer to use steps rather than a ramp. Steps allow people to rest with their feet on a level surface, rather than a sloping one that can cause pain in the ankles for older people or those with arthritis. Some visually impaired people also prefer to use steps because they give reference points, which can assist with mobility and orientation.

- Ramps with a gradient of 1 in 20 or steeper should always be accompanied by steps.
- Travel along ramps can become tiring and so no section of ramp should exceed 10 m in length or 500 mm in rise.

Figure 4.16 Well-designed ramps can increase access to existing buildings.

- A series of ramps to a building should not rise in total more than 2 m. Beyond this rise an alternative means of vertical travel, such as a lift, should be considered.

Gradient   A ramped approach should have the lowest practical gradient:

- 1 in 20 is considered desirable;
- 1 in 15 is acceptable;
- 1 in 12 is the absolute maximum. A gradient of 1 in 12 may be too steep for some wheelchair users and will prevent them from accessing a building. Egress on a 1 in 12 gradient may be difficult to achieve safely.

BS 8300:2001 and Part M of the Building Regulations give guidance on ramp gradients.

Width   The minimum surface width of a ramp, as given by BS 8300:2001, is 1200 mm.

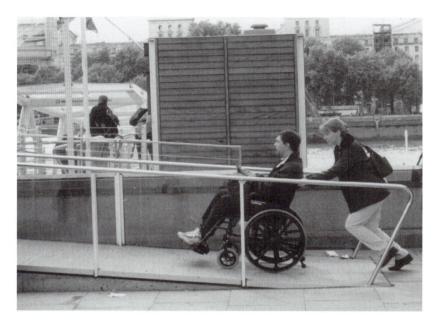

Figure 4.17 A ramp with a gradient of 1 in 12 may be too steep for people to use easily and conveniently.

- If the width is less than 1800 mm, two wheelchair users will not be able to pass. Where this occurs there should be a clear unobstructed view along the length of the ramp to ensure that wheelchair users know they can complete their travel without having to reverse up or down the slope.
- Ramps of the minimum width should not exceed 5 m in length, and even then they are not suitable for heavy traffic. Passing places should be incorporated at landings if it is not possible to provide a wider ramp.

Reversing up or down a slope, for whatever reason, is a dangerous, often impossible, manoeuvre for a wheelchair user. The need to do it should be avoided in all possible circumstances – and it can be with careful thought and good design.

Landings

- Resting places, in the form of level landings, should be provided along the length of the ramp as follows: every 5 m for a gradient of 1 in 15 and every 10 m for a gradient of 1 in 20. Ramps with a gradient of 1 in 12 should not exceed 2 m in length.
- Intermediate landings should be at least 1500 mm long. Landings of 1800 mm by 1800 mm could also act as passing places.
- Level landings should also be provided at the top and bottom of each ramp, and at changes of direction. The landings at the top and bottom should be a minimum of 1200 mm long and clear of any door swing.
- Unless under cover, a landing should have a cross fall not exceeding 1 in 50 to help drain surface water. If kerbs or upstands prevent water run-off, drainage holes should be provided. Care should be taken to ensure that their size and position do not endanger ramp users. Slots should run at right angles to the direction of traffic.

Surfaces

- Surfaces should be smooth, firm, slip resistant even when wet, and easy to maintain.
- To alert people with visual impairments to the presence of a ramp, the landing areas should contrast in colour and luminance with the sloping sections.
- If different surfaces are used for ramps, landings and approach paths, it is important that the coefficients of friction are similar to minimise risk of stumbling.

A tactile warning surface (corduroy pattern) should not be used at the top or bottom of ramps. This surface is intended for use at steps and may cause confusion, or even be potentially dangerous, if used incorrectly.

Handrails   Usually wheelchair users will not require a handrail on a ramp. However, in adverse weather conditions or if the ramp is long and steep, handrails may assist wheelchair users in steadying themselves. For ambulant disabled people, especially those with less strength on one side, a handrail is an essential means of support when going up or down a ramp. For visually impaired users, a handrail may be used as a tactile guide that, if extended beyond the end of the ramp, can give useful information about the beginning and the end of the ramp itself.

- Appropriately designed handrails should be provided to both sides of any ramp, with the top edge at a height of 900 mm to 1000 mm above the ramp surface.
- On wide ramps, exceeding 2 m, a central continuous handrail can be provided in addition to side handrails.
- Handrails should be able to be gripped and should extend beyond the start and finish of the ramp by at least 300 m.

For detailed guidance on handrail design see *Internal steps and stairs.*

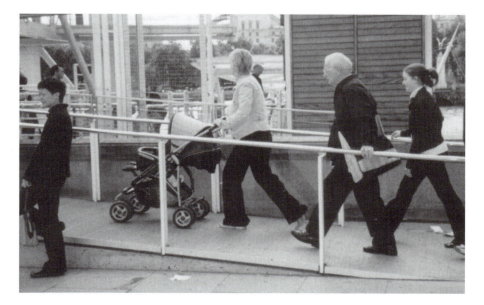

Figure 4.18 All ramps should have appropriately designed handrails.

Guardrails and kerbs   Any guardrails or balustrades should be designed with the safety of all users in mind. There should be no risk of wheelchair users catching their feet between balusters. If glazed panels are used, they should be fit for the purpose.

- Where there is no handrail to the open side of a ramp, there should be a raised kerb of at least 100 mm, differentiated from the ramp by colour or luminance. This will aid wheelchair users and act as a tapping rail for people using canes.

Portable or temporary ramps   Portable or temporary ramps should not be used as a design solution in new buildings. Where it is not possible to install a permanent ramp in an existing building, for example, within a listed building, the use of a portable or temporary ramp may be necessary. Such a ramp should be designed and used so that it does not constitute a hazard for any users of the building or environment.

The use of a temporary ramp to provide access may be seen as reasonable provision under the DDA in some circumstances. Factors to be taken into account may include the size of the company, the physical practicalities of providing a permanent ramp, the cost and whether the employment opportunity or service could be provided by a reasonable alternative method. There will also be management issues relating to the use of the ramp such as storage, responsibilities for setting up the ramp and offering assistance to users, maintenance and procedures for requesting use, all of which must be dealt with if the ramp is to provide reasonable access.

- A portable or temporary ramp should be well contrasted with its background and well illuminated.
- The surface width of the ramp should be at least 800 mm.
- The surface should be slip resistant and well drained.

- There should be upstands to prevent wheels slipping over the edges of the ramp.
- Gradient should not exceed 1 in 12.

Assistance should always be available for people wishing to use a portable ramp. It is potentially very dangerous to leave a portable ramp in position, at a steep gradient, and allow people to use it without assistance.

Pavement realignment   Where there is insufficient space to provide a ramp at the entrance to an existing building, it may be possible to ramp a section of pavement to overcome the change of level. Factors to be considered will include the location of other entrances, the width of the pavement and the resulting height of the kerb. Collaboration with the relevant authorities would be required.

Figure 4.19 Gently ramping a section of pavement can overcome a change of level at an entrance.

External steps and stairs

Some ambulant disabled people will find steps easier to negotiate than ramps. The profile of the steps, the surface finishes and the provision of suitably designed handrails is critical.

- Steps should always be provided as an alternative to ramps.
- Handrails should be provided, however short the flight.
- Isolated single steps should be avoided where possible.
- Individual flights should not contain more than 12 risers.
- The width of a stepped access route should be at least 1000 mm.

For detailed guidance on steps and stairs see *Internal steps and stairs.*

Handrails   For many people, handrails are an essential source of information and support. They can warn of the presence of steps, guide and support people using the steps and provide information

Figure 4.20 Tapered risers can present a serious tripping hazard and should be avoided.

on the start and finish of a flight and the level reached. The provision of well-designed and positioned handrails is essential for safe, independent use of steps.

Handrails should always be provided to each side of a flight of steps. People may be weaker on one side and require a handrail for support. The division of wider flights into separate channels will allow easier access to handrails when many people are using the stair.

The horizontal extension of a handrail beyond the first and last steps allows an individual to steady or to brace him or herself before ascending or descending, provides support to ascend the final riser and will signal the start or finish of the flight to people with visual impairments.

For detailed guidance on handrail design see *Internal steps and stairs*.

Steps with handrails to only one side are difficult or impossible for some disabled people to use. Individuals who have the full use of only one side of their body will need the assistance of a handrail when both ascending and descending the stairs. Guide dog users may not be able to change the side the guide dog operates to allow use of a handrail; people with visual impairments who use a cane for mobility may have similar restrictions.

Equally important are the 300 mm extensions to the top and bottom of handrails, continuous handrails around landings and tactile information to advise people of floor levels. These elements should not be seen as optional – for some people they are the only way that they can obtain information about their environment to allow them to move around independently, comfortably and safely.

Tactile warning   There can be a risk of tripping or losing balance, particularly at the head of a flight of steps, and so a warning surface is recommended to alert people to the imminent change of level.

Figure 4.21 The horizontal extension of a handrail can provide support and also information on the start and finish of the flight of stairs.

- It is recommended that a tactile warning surface (corduroy pattern) be provided at the top and bottom of each flight. The surface should extend beyond the width of the flight if practicable. See Guidance on the Use of Tactile Paving Surfaces (DETR 1998).

Entrances

The entrance to a building will have a significant impact on both the perceived and actual accessibility of the building. The principal entrance should be easily found and used and allow entry by everyone. People should not be segregated on entry by the provision of a separate entrance for those who cannot use the main entrance. There may be existing buildings where an entrance cannot be made accessible and so an alternative means of entry is required, but this approach should be avoided wherever possible and never used in new buildings.

Entrance design

- The principal entrance to a building should allow access by everyone.
- Entrances should be visible on approach and be able to be distinguished from the façade of the building.
- The location of entrances should relate to external routes and car parking.
- Weather protection should be provided, such as a canopy or recess, unless there are freely accessible automatic doors.
- There should be sufficient manoeuvring space internally and externally to allow everyone, but especially wheelchair users and guide dog users, to correctly position themselves to approach and enter the building.
- The lighting levels on entry into a building should be graduated to allow adjustment from a bright exterior to a lower internal light level.

Entrance doors

Doors should allow easy entry and egress. Factors to be taken into account include the minimum clear opening width of the door, the resistance that needs to be overcome to open the door, the opening mechanism and the provision of vision panels.

- Entrance doors should give a minimum clear opening width of 800 mm through one leaf. The opening width should be increased to accommodate any projections such as full-height pulls or a weatherboard.

> Where there are double doors, at least one leaf should be of sufficient width to allow entry (800 mm minimum), as people may not be able to open both leaves simultaneously.
>
> In some cases, for example, where applications for funding are being made to bodies with specific access requirements, a minimum clear opening width of 900 mm may be required.

- Single doors should have at least 300 mm space beside the leading edge of the door to allow anyone with limited mobility to approach and open the door.
- Door closers should be avoided if possible, but where necessary should be adjusted to the minimum force necessary, be slow in operation and regularly maintained. The maximum closing force at the leading edge of the door should not exceed 20 N and should be exerted between 0 and 15° of final closure. Delayed action closers are preferred and should be fitted where possible.
- Outward opening doors should be recessed or the swing area adequately protected to prevent collisions.
- Vision panels should be provided in doors in frequent use. The minimum zone of visibility should be between 500 mm and 1500 mm above the floor.
- Glazed doors and side panels may require manifestation to increase visibility. Manifestation should be well contrasted against its background and highly visible at all times of day, and where appropriate, in both natural and artificial light.

The provision of good, clear manifestation at design stage will remove the possibility of *ad hoc* and possibly unattractive additions later on.

The design of manifestation can offer an excellent opportunity to promote the company or service at the very place most people will be looking when they enter a building or move around it.

- Where glass doors are part of a glazed screen there should be a contrasting frame or other means of differentiating doors and screen.
- Edges of glazed doors should be clearly visible when the doors are in an open position.
- Door furniture should be distinguishable, in terms of colour and luminance contrast, from the door and be positioned where it can be easily reached, gripped and used with minimum effort.

Figure 4.22 Manifestation should be able to be clearly seen in all lighting conditions. It can also be used to help identify entrances.

Full-height door handles and other features, such as weatherboards, must not restrict the clear opening width of doors.

- Lever-style handles allow use by elbows or the edge of the hand. A return at the end will prevent the hand from slipping off the handle and help prevent clothing being caught.
- A kicking plate can protect a door from damage caused by wheelchair footrests if full width and at least 400 mm deep.

Automatic doors

Automatic doors provide good, easy access and will benefit all building users. Sliding doors are preferable in areas of heavy traffic. Where swing doors are used, care should be taken to protect the door swing area. Advice on the presence and operating style of automatic doors should be provided on approach, using signs; a change in floor surface can also be used.

- Activation can be automatic or manual. Where there are manual activators, they should be clearly signed, identifiable (perhaps by the use of good colour contrast) and positioned and designed to allow easy approach and use.
- Swing doors require protection to swing area, particularly if opening towards users or into a circulation area.
- Low-energy swing-door operators do not require additional safety equipment such as presence sensors if they have low opening forces, low opening speed and obstacle default function.
- Sliding doors generally provide very good access and may be preferred in areas of heavy traffic.
- Doors should remain open for long enough to allow a slow moving person to pass through.

Figure 4.23 Where there is a choice of doors, the majority of people prefer to use the automatic door.

Figure 4.24 Automatic swing doors should be clearly signed and the swing area protected.

- All automatic doors require safety devices to ensure that the door does not close if there is an obstruction. With the exception of low-energy systems, these devices should use photocell methods to ensure closing does not begin until doorway is clear.
- Sliding/folding doors are useful where space is restricted and often can be used in existing door openings.

Automatic doors generally offer very good access for disabled people, but doors that swing towards the user can be dangerous and off-putting. Tactile information could be provided by the use of a different floor surface in front of the door to warn of the direction of the door swing.

Revolving doors

Revolving doors do not provide good access for everyone. Even the larger automatic revolving doors, sometimes found at shopping centres or supermarkets, are not recommended as many people find them difficult or uncomfortable to use. Wheelchair users, people using mobility aids such as sticks or crutches, guide dog users and older people may not be able, or may not have the confidence, to use any size or type of revolving door.

A revolving door at the entrance to a building will not give a message of accessibility and openness, and, importantly, it will cause visitors to be segregated on entry to the building.

Where they are used, there must be an alternative entrance with a clear opening width of 800 mm, immediately adjacent and clearly visible. This should be a swing door, preferably automated and always operational.

Where swing doors are provided as an alternative entrance, they should be operational at all times when the building is open and the revolving door is in use. If the alternative entrance is kept locked, the message given to visitors is 'you can come in if you can use the revolving door, otherwise you will have to wait for someone to come and let you in'. Many people will see this arrangement as discriminatory.

Where climate control is a concern two sets of automatic sliding doors can be used, offset, with a lobby between.

Thresholds

Thresholds should be flush wherever possible. Opening a door whilst simultaneously negotiating a raised threshold can be extremely difficult for some users and any change in level can present a tripping hazard.

There need not be a conflict between user needs and water-proofing, with careful design and management both sets of needs can be met.

- Thresholds should be flush wherever possible.
- Where a raised threshold is unavoidable, the maximum change in level should be 15 mm, providing the raised section is clearly visible and the floor finish graded to provide a flush finish. However, some people with ambulant disabilities will find any changes in level difficult, if not impossible, to negotiate.

Doormats

Inside entrances a floor surface should be provided that will remove rainwater from shoes and wheels to prevent floor finishes becoming slippery. Where a doormat is provided, it should be firm and flush to allow easy wheeled passage.

- Coir and other deep pile materials are not suitable for wheeled passage.
- Mats that are not properly recessed can cause tripping.

Figure 4.25 Good use of fixed mats – but loose mats represent a serious tripping hazard.

Any fitted doormats should be flush with the floor, firm and close fitting to the mat well.

Lobbies

Lobbies should be of sufficient size to allow easy access, egress and manoeuvre. The layout and the door design should allow all users, including wheelchair users, ambulant disabled people, guide dog users and people with children or pushchairs, to open doors independently and move clear of one door before negotiating a second.

Entry systems

All entry systems should be designed and located to allow easy approach and use. Security systems should suit the requirements of all people entering a building, including wheelchair users. Turnstiles or barriers that segregate people on entry to a building should not be used.

- Entry systems should be suitable to meet the needs of all users, including those with limited dexterity, visual and speech impairments and those users who are deaf or hard of hearing.
- Entry phones should be clearly identifiable (perhaps with the use of colour and luminance contrast) and should be placed at a height that can be operated from a standing or seated position. The maximum height of the highest operating button should not be more than 1200 mm above finished floor level.
- The design of press keypads should be suitable to allow operation by people with restricted dexterity and incorporate sufficient contrast to be seen. They should also be set at a height that can be used from either a standing or seated position.
- If a control device contains an audible communication system, perhaps to announce arrival or to seek assistance, the needs of

Figure 4.26 Systems used to control access and for communication must be accessible, and usable, for everyone.

people who are deaf or hard of hearing must be considered, perhaps by incorporating an induction coupler and an LED display.

- Clear, unambiguous information on what disabled people should do to attract attention or summon assistance, if required, should be displayed on or near the control device. Any contact phone numbers that could be used by mobile phone users to call assistance if they cannot exit their vehicles should be clearly displayed on or adjacent to the entry control device.

- Management responses to dealing with someone who requires assistance to enter a building or other environment should be written down and practised to ensure they are operable at all times. Staff should also be fully trained in dealing with the communication needs of people who are deaf or hard of hearing.

Figure 4.27 This car park control requires reach, dexterity, speech, hearing, and a right arm. It is possible to hold a driver's licence without one or more of the above – but you need all of them if you wish to enter this car park. Is this reasonable provision?

If an entry system is being used to control entry to a building in which an employment opportunity or service is being provided, there may be duty under the DDA not to discriminate in allowing access to the building. It may be possible to replace an unsatisfactory system with an alternative – perhaps one incorporating CCTV – or to use appropriate management techniques and staff training until the system can be replaced, but the issue cannot be ignored.

Exits

Exit doors, particularly those used for emergency egress, should incorporate the features recommended for entrances as appropriate.

Figure 4.28 Management must consider how to enable disabled people to escape in an emergency – taking account of assistance requirements especially if lifting is necessary.

Reception areas

On entering a building the location of the reception point should be obvious and logically placed. Reception areas should be well lit, routes should be clearly defined and unobstructed, and there should be clear information on facilities available within the building.

- The waiting area should be quiet and well lit to allow easier communication, especially for people with hearing impairments.
- There should be clear unobstructed routes with sufficient space for manoeuvring.
- All furniture should be adequately contrasted in terms of colour and luminance contrast with the background against which it will be viewed. In smaller areas, this may simply be the walls, in larger areas it may be the wall, floor or other furniture.

- If there is an audible communication system, it should be supplemented by visual information.
- Where there is likely to be a high level of background noise, an induction loop should be provided and its presence indicated by the standard symbol (see also *Reception desks*).

Fully trained staff at key points, such as car park entrances, reception and help desks, can improve considerably the perceived and actual accesssibility of a building or facility whilst also addressing many security and safety issues.

Staff should be trained in understanding the needs of disabled people and how to interact with them. Staff should know of any potential dangers or hazards that may affect disabled people using the environment and be aware of what to do about it if required. Where important facilities, such as accessible toilets, are located should be known to them, together with information on the easiest routes of how to get there.

Key staff should also understand the needs of visually impaired users and undergo basic mobility training to assist visually impaired users to move around an environment safely if required. They should also have some basic training in clear speaking and perhaps basic sign language, to assist those users who are deaf or hard of hearing.

Seating should be made available where there is a likelihood of waiting being necessary. To ensure that the needs of as many users as possible are met, a selection of seating should be provided, with different seat heights and with and without arms. If fixed seating is provided, there should be some movable seating to accommodate wheelchair users sitting with other people.

- Where seating is arranged in rows, there should be clear space of 900 mm minimum width (1200 mm preferred) in front of a row of seats to allow people to pass along a row to a seating position.

- A mixture of fixed and removable seats should be provided to accommodate different seating layouts and numbers of disabled people.
- Integrated space for wheelchairs should be provided to allow wheelchair users to sit alongside seated companions.
- Space adjacent to seating should be provided for working dogs.
- Seats should be of sufficient size to accommodate larger people and some should have firm seats and armrests that are sturdy enough to assist users when sitting or standing.

Reception desks

The reception desk should be easily seen and appropriately signed. There should be sufficient manoeuvring space to allow people to approach and use the desk. Communication at the desk should be possible at seated and standing heights.

The design of a reception desk, and particularly the requirement for a low section for the benefit of wheelchair users, is dependent upon the way in which the desk will be used. If no paperwork needs to be completed at the desk, a lowered section may be desirable, but is not essential provided communication is possible when seated. However, if people visiting the building are required to sign in, or fill in other paperwork, then some means of doing this must be provided.

Where there is a lowered section, there should also be a knee recess to allow a wheelchair user to pull up close enough to use the desk. It may be possible to alter an existing desk or to add a drop-down shelf that could be lowered when required.

In existing buildings, looking at the way a desk is actually used can allow a decision to be made on whether immediate replacement or alteration is necessary, or whether this can wait until the desk is due to be replaced or refurbished.

- Access for wheelchair users should be possible to both staff and visitor sides of reception desks and counters.
- A low-height counter or lowered section should be provided where it is necessary for communication or signing-in. Seating should be provided at low-height counters.
- The receptionist's face should be clearly visible and well lit to allow lip reading. If there is a glazed screen, it should be able to be opened to allow for direct communication. If it cannot be opened, or if there is likely to be background noise, an induction loop should be provided for the benefit of hearing aid users.

Glass partitions should be avoided because they make lip reading difficult to deaf or hard-of-hearing users. If glass partitions are unavoidable, non-reflective glazing and a speech enhancing system should be used.

- Induction loops are often fitted at reception desks and can be used by people with a hearing aid equipped with a T switch. This is estimated to be about 1.5 million of the 8.5 million people in the UK who have a hearing impairment or loss. For the others, good communication by key staff and other methods of assisting people who are deaf or hard of hearing must be considered. Good lighting design, appropriate decoration to surfaces, staff speaking clearly, information being available in alternative formats (for example, 14 point sans serif font on non-reflective yellow paper) and, if possible, some awareness of or training in British Sign Language are all important.

The provision of an induction loop alone does not mean that an employer or service provider has met the needs of the deaf and hard-of-hearing population.

Figure 4.29 This desk has a low section – but the writing shelf and the lack of knee space would prevent easy use by a wheelchair user.

Horizontal circulation

Ease of navigation

A rational and simple building layout is easier to navigate and remember. The entrance to a building should be located clearly and in an obvious position. The reception area should be close to the entrance and give easy access to circulation routes throughout the building. Circulation areas should be clear of hazards and use colour and contrast to identify routes, layouts and obstacles.

Clear directional information should be provided with resting places on longer routes and conveniently located and clearly signed lifts where required. A simple, well-planned circulation layout will encourage easy orientation.

Circulation routes through open-plan spaces should have sufficient clear width, which should be maintained when layouts

Figure 4.30 Route to the reception desk is identified, however, colour and luminance contrast would have made this information available to a greater number of people.

are altered. It can be helpful to define routes using colour, tonal or textural contrast.

Corridors and passageways

Corridor widths should accommodate wheelchair manoeuvring and passing, guide dog use and, where relevant, the needs of people with pushchairs.

- A minimum width of 1200 mm is recommended for corridors and other circulation routes. This width is essential at door-opening positions to allow a wheelchair user to turn through 90°. Wider corridors will be required where there is heavy use.

- In buildings where there are likely to be significant numbers of wheelchair users the corridors should be at least 1800 mm wide, or have passing places at reasonable intervals.
- Where corridor width is restricted, for example in existing buildings, wider door openings should be considered and passing places provided.
- Corridor widths should be unobstructed with wall-mounted fittings recessed wherever possible.
- Outward opening doors can be hazardous, particularly to people with sight impairments, and should be recessed.
- Splayed or rounded corners can ease circulation along corridors.

Surfaces

The choice of floor and wall surfaces can greatly affect the ease of use of an environment. Hard surfaces can cause sound reverberation and high levels of background noise, which can cause difficulties in communication. Thick pile carpets will hinder wheelchair passage. Shiny floors can appear slippery and are not liked by visually impaired and older people who feel unsafe when walking on them. Heavily patterned floor finishes can be confusing to people with visual impairments. Adequate colour and luminance contrast are essential to define surfaces and objects placed upon them.

- Highly reflective finishes should be avoided for floor, walls, doors and ceilings.
- Colour and luminance contrast will help define floor, wall, ceiling and door surfaces and objects placed upon them.
- Floor surfaces should be firm and flush.
- Floors should be slip resistant, but also not look as if they might be slippery.
- Junctions between different floor finishes should be detailed to prevent tripping and allow easy wheeled passage.
- Where carpet is laid, it should be firmly fixed, non-directional and have a shallow dense pile.
- Surface-laid rugs or mats must be avoided in public buildings, as they present tripping or slipping hazards.

Figure 4.31 How many doors are there? Reflective finishes can be very confusing for someone with poor vision.

Figure 4.32 Many people would find this floor finish disconcerting, but especially those with poor vision, older people and people using mobility aids such as crutches. It may not be the selection of the floor covering that is the problem, but the way it is being cleaned and the impact of daylight and artificial lighting.

- The choice of materials for floor, wall and ceiling surfaces should give an acoustic environment that allows audible information to be heard and aids orientation.
- Textured surfaces can be used to impart information to people with visual impairments. It is important to use recognised textures or to provide a key.

Glazed walls and screens

- Glazed walls and screens that are adjacent to doors, or form part of an enclosure, should have clear manifestation, each element at least 150 mm square, which contrasts with the background in all lighting conditions. The manifestation should be placed at standing and seated eye level, 1500 mm and 1050 mm above floor level.

Figure 4.33 Are the squares manifestation on the glass or decoration on the train? Critical information is lost – even for those with good vision.

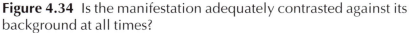

Figure 4.34 Is the manifestation adequately contrasted against its background at all times?

Handrails

Many older people and people with visual impairments will benefit from the provision of handrails along corridors to assist with physical support or give tactile information.

- Where handrails are provided, they should be designed to be gripped, be well contrasted with the supporting wall and be as continuous as possible.

For detailed guidance on handrail design see *Internal steps and stairs*.

Internal doors

Doors on circulation routes will be less of a barrier if they are light enough to open easily, give the required clear width through one leaf and have handles that can be used with minimum effort.

Door width and location

- Internal doors should have a minimum clear opening width of 750 mm, though 800 mm is strongly recommended.
- The minimum opening width must take into account any door furniture and doorstops.

> The layout of furniture and fittings, in a building in use, may affect the clear opening width of a door.

- Double doors should provide the minimum required opening width of 750 mm (800 mm preferred) through one leaf.

> Where overall width is restricted, double doors can have leaves of unequal size to give the necessary clear opening width through one leaf.

- There should be sufficient space around doors to allow users to manoeuvre and approach. A minimum space of 300 mm should be provided adjacent to the leading edge of a door that swings towards the user to allow easy independent use.
- Doors opening into a circulation route should be recessed or have the swing area protected.

Door furniture

- Door furniture should be distinguishable, in terms of colour and luminance contrast, from the door and be designed and positioned to be easily reached, gripped and used with minimum effort.
- Lever-style handles allow use by elbows or the edge of the hand. A return at the end will prevent the hand from slipping off the handle and helps prevent clothing being caught.
- A kicking plate can protect a door from damage caused by wheelchair footrests if full width and at least 400 mm deep.

Figure 4.35 A good style of door handle is essential to many disabled people.

The use of appropriately contrasted door handles, fingerplates, hinges and kicking plates can assist people with visual impairments to recognise the presence of a door, even if the door is the same colour as the wall surrounding it. There are very few things in the built environment that mimic the visual appearance of the pattern of door furniture. The colour and shape of door furniture can be indicators not only of the presence of a door, but also of the way it opens.

This type of information, often giving a person with a visual impairment hints about things they may well have once seen in more detail, can help enormously when someone is moving through a building and searching for features. Similarly, the careful use of subtle shadows, such as those around the panels on a panelled door, can give very useful clues to the presence of a door, even if there is little colour contrast between the door and its surround.

Eighty-two per cent of people with impaired vision actually see something; their ability to use that residual vision to gather information about their environment can be maximised by careful design.

Kicking plates should either be glued to the door or fixed with countersunk screws. Projecting screw heads may cause injury to a disabled person and/or damage to the wheelchair.

Door closing   Door closers should be adjusted to the minimum force necessary, be slow in operation and regularly maintained. The closing force of a single swing door, which is not required for fire protection, should not exceed 20 N. The maximum force should be exerted between 0 and 15° of final closure. Delayed action closers are preferred and should be fitted wherever possible.

- The closing force on the leading edge of a double swing door leaf, which can be pushed open in both directions, should not exceed 30 N.
- Where a door has no closing device it is useful to fit a pull handle to help people close the door behind them.

Fire doors   Fire doors fitted with self-closing devices can prevent easy circulation around a building. It is important that doors and closers are specified that will allow easy use, while maintaining the required level of fire protection.

- Where the closing force at the leading edge of a door on a circulation route exceeds 20 N, an electrically powered hold-open device, conforming to the requirements of BS EN 1155, should be installed. A device of this type can hold a swing door at a fixed position or allow it to swing freely. Door closing can be activated by a signal from the fire alarm system, manually or by fail safety operation.
- The closing force on doors with a push action in one direction only, for example final exit doors into a protected stairway, should not exceed 30 N.
- If smoke seals are required, installing an angle seal as an independent item in the doorframe can reduce the force required to open the door.

Even when a door is held open, the force required to open it from a closed position should still be checked when carrying out an audit. When the hold-open device is deactivated in an emergency, it is essential to ensure that the door can be opened easily to allow people to escape or that there is a management procedure in place to deal with such an eventuality. What must be avoided is the possibility of the door trapping people when an emergency occurs.

Visibility

- Colour and luminance contrast will allow doors and/or doorframes to be distinguished from their surroundings.
- Glazed doors and side panels should be clearly defined with manifestation, each element at least 150 mm square, at 1050 mm and 1500 mm above the floor to increase visibility.

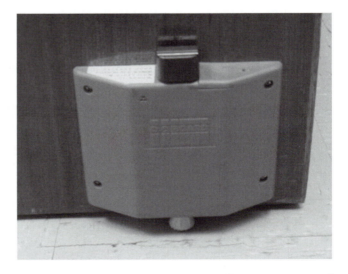

Figure 4.36 Fire doors can be held open to allow easier circulation.

Manifestation should be well contrasted against its background and highly visible at all times of day and in all lighting conditions.

• Fully glazed doors that form part of a glass screen should be clearly differentiated from the screen.

Sometimes, designers and building managers are unsure whether it is necessary to contrast the door with the surrounding wall or if simply contrasting the architrave with the wall, and having the door the same colour as the wall, is adequate.

Most visually impaired people have a strong preference for the whole door being contrasted with the surrounding wall. Contrasting only the architrave will still make most people aware of the possible presence of a door, but it will take longer for them to be sure they are seeing a door as opposed to some other feature.

Vision panels   Doors on circulation routes should have vision panels. These are also useful on other doors to allow people to determine whether a room is in use.

• Vision panels should provide a minimum zone of visibility of 500 mm to 1500 mm above floor level.

Automatic opening   Automatic or power-assisted opening devices will be of benefit to many users and some can be fitted to existing doors. Activation can be automatic or manual. Where there are manual activators they should be clearly signed, positioned and designed to allow easy approach and use (see also *Entrance doors*).

Figure 4.37 Manual activators should be designed and located to be easily identified and used.

Vision panels can be useful for people who are deaf or hard of hearing to help them gain the sort of information that others might gather audibly. For example, if someone knocks on a door seeking entry into a room, they may be unable to hear a response of 'come in' or 'please wait'. The only way that they will know if their knock has been heard, or indeed if the room is occupied, will be to open the door and look, which may not be appropriate. Vision panels can help to prevent this problem.

Vision panels can also save time when checking evacuation from a building in an emergency – good building management and accessibility working hand in hand.

Lobbies   Internal lobbies should be avoided wherever possible. If unavoidable, they should be of sufficient size to allow easy access, egress and manoeuvre. The layout and door design should allow all users, including wheelchair users, ambulant disabled people, guide dog users and people with children or pushchairs, to open doors independently and move clear of one door before negotiating a second.

Vertical circulation

Stairs, steps, ramps and lifts all provide means of vertical movement within a building. All should be designed to allow easy and safe use.

- Ramps and/or lifts should be provided at all changes of level for people who cannot, or prefer not to, use stairs.
- All internal changes of level, including single steps and short ramps, should be clearly indicated to reduce the risk of tripping or losing balance.

Internal steps and stairs

Stairs are probably the most common means of achieving change of level and should be designed to accommodate safe and easy use by everyone. Lack of suitable handrails for support, projecting nosings or open risers that can trap toes, and lack of warning on approach can result in stairs that are difficult for people to use and potentially hazardous.

- Handrails should always be provided, however short the flight.
- Isolated single steps should be avoided where possible.
- The unobstructed width of stairs should be at least 1000 mm.
- A level landing of at least 1200 mm should be provided at the top and bottom of each flight of stairs.
- The preferred maximum rise between landings should be no more than 12 risers. (In certain small premises a maximum of 16 risers may be acceptable.)

Risers and goings

- The height of the riser should be between 150 mm and 170 mm. The going should be 250 mm to 300 mm, with a preference for 300 mm.

Figure 4.38 Open risers should always be avoided; they can be visually confusing and catch toes, sticks, callipers, etc.

- Open risers and deeply recessed risers should not be used as they can catch toes, sticks, callipers, etc. Open risers can also be visually confusing for people with sight impairments.
- The nosing of each step in a flight of stairs should be adequately contrasted with the remainder of the step and the floor covering adjacent to the top and bottom of the flight. Contrast should be provided by the use of a single strip across and around the nosing, extending 40 mm on both the riser and the going, with adequate colour and luminance contrast. Patterned highlighting, especially 'sharks tooth', should not be used as it can give a confusing message, which can be potentially dangerous.
- Steps without projecting nosings are preferred. If there is an overlap it should be less than 25 mm.
- Tapered treads and spiral stairs are not recommended for use in public buildings.

Figure 4.39 Each step nosing should contrast in colour and luminance with the remainder of the tread and should be clearly visible when ascending and descending stairs.

Steps with goings that are less than the recommended minimum length can be difficult or impossible for some people to use. If someone is unable to place his or her whole foot on the step, all the weight will be supported on the sole of the foot when ascending the steps and on the heel when descending, the only alternative being to turn the foot sideways to obtain the maximum area of support.

All these manoeuvres would be extremely difficult for many disabled and older people to perform and this may result in them not being able to use the stairs. There may also be consequences relating to the DDA where a physical feature makes it impossible or unreasonably difficult for disabled people to access a service or employment opportunity.

Figure 4.40 The area below overhanging stairs should be guarded.

The design of a stair should not present a danger to users of a building, either when using it or moving past it. Importantly, the underside of staircases should be protected to prevent all users, but especially those with a visual impairment, colliding with the structure of the stair.

Under-stair areas with low headroom should be protected with a fixed rail, houseplants or other feature to avoid the possibility of users, especially those with a visual impairment, colliding with the staircase.

Handrails Handrails are used for a variety of reasons when ascending or descending a stair, and the design and provision of them should take varying needs of users into account wherever possible. For example, people with impaired vision often rely heavily on handrails to orientate themselves on staircases, to determine when they have reached the top or bottom of a flight of steps or in establishing a change of direction. All people will need a handrail that is easy to grip if they trip or fall on the stairs.

Therefore, it is important that a handrail is visible, reachable, is strong enough to provide physical support if needed and offers good tactile information about the stair to the user, both when ascending or descending the stairs.

Except in existing buildings, where narrow staircases and other safety issues may prevent it, handrails should always be provided to each side of a flight of steps. People may be weaker on one side and require a handrail for support. (It should be noted that the installation of a platform or stair lift might prevent the use of one handrail.) The division of wider flights into separate channels will allow easier access to handrails when many people are using the stair.

The horizontal extension of a handrail beyond the first and last steps allows an individual to steady or to brace him- or herself before ascending or descending, provides support to ascend the final riser and will signal the start or finish of the flight to people with visual impairments.

- A handrail, properly contrasted (in terms of colour and luminance) against its background, should always be provided to both sides, however short the flight.
- Handrails must be of a size and shape that is easy to grip and of circular or oval profile. A circular handrail should have a diameter between 40 mm and 50 mm and the preferred dimensions for a handrail with an oval profile are 50 mm wide and 38 mm deep. The oval profile should have rounded edges with a radius of at least 15 mm.

A non-circular handrail with a flat upper surface gives better hand and forearm support. However, it must be able to be gripped.

- When fixed to, or placed adjacent to a wall, a clear space of between 50 mm and 60 mm should be provided between the handrail and the wall surface. To minimise the risk of supports to the handrail injuring a user's hand or reducing the ability to grip the rail, a clearance of at least 50 mm should be provided between the handrail and any cranked support or balustrade.

- Handrails should be continuous to stairs and landings, be able to be gripped along their full length and extend horizontally at least 300 mm beyond the first and last nosing in the flight.
- The vertical height to the top of a handrail should be between 900 mm and 1000 mm from the pitch line of a flight of steps, and between 900 mm and 1100 mm from the surface of a landing.
- A central handrail should be provided on wide flights.
- The handrail should either return to the wall or have a positive end.
- Nylon and timber handrails are pleasant to the touch and are best suited for people who may be affected by the contact temperature of a handrail. Mild steel handrails will tend to be cold to the touch.

> Tactile warning studs or markers on the handrail can be a very useful indicator of the beginning or end of flight and to indicate the floor level reached.

Figure 4.41 Contrasting nosings and tactile warning can help people use stairs.

Hazard warning   There can be a risk of tripping or losing balance particularly at the head of a flight of steps and so a warning surface is recommended to alert people to the change of level.

- A tactile warning surface (corduroy pattern) can be provided at the top and bottom of each flight. The surface should extend beyond the width of the flight if practicable. If the warning surface is not used there should be a change of colour and texture in the floor finish on approach to the stair.
- Where there is access beneath a stair, the underside should be protected to prevent people from colliding with the structure.

> Care should be taken to ensure that stairs are not a continuation of the normal line of pedestrian travel, for example along a corridor, without any warning given of their presence.

Internal ramps

Ramps should be used where it is necessary to address unavoidable changes in level and not simply to overcome changes of level that could have been avoided by good, thoughtful design. The presence of a ramp should be clearly indicated to warn people of its presence and prevent tripping. Colour or luminance contrast in floor finish can be used.

Some people may have difficulties walking on a sloping surface and will prefer to use steps rather than a ramp. Wherever ramps are provided steps should accompany them.

- Ramps in internal corridors should be highlighted to increase visibility.
- Ramps should always have suitably designed handrails.

With the exception of those issues relating specifically to the external environment, the design considerations for internal ramps are as those for external ramped access (see *External ramps*).

Lighting on stairs and ramps

Good lighting is essential on stairs and ramps, though bright directional lighting, glare or strong shadows can be disorientating and cause accidents.

- Illuminance at tread/floor level should be at least 200 lux.
- Lighting incorporated into stair risers is likely to be disorientating to all users and should not be used.

> Carefully positioned lighting can be used to help identify the location of risers and goings in stairs. Contrasting nosings are always recommended to identify steps, but if they are not present and cannot be added, for example in some historic properties, good lighting can help.

Escalators

Escalators can be useful where there is a need to move large numbers of people between floors, but can present a barrier to many people. They should be carefully designed to allow safe use and there should always be an alternative means of access provided for wheelchair users, people with pushchairs, guide dog users or any other people who do not wish to use the escalator.

- The entry and exit to the escalator should be clearly visible, well lit and free from unnecessary obstructions.
- There is no requirement in regulations to provide a tactile floor surface on the approach to an escalator. However, it is useful to provide some contrast in the floor covering at the top and bottom landing to alert people to the presence of an escalator.
- The direction of travel should be clearly indicated, possibly by a red or green light. This is of particular importance where escalator direction is altered, for example during busy periods. The indicator should be located and designed to be clearly visible on approach to the escalator.

- Audible announcements can be used to warn of approaching end of escalator.
- Handrails should have colour and luminance contrast and extend at least 150 mm beyond the entry and exit points.
- Handrails should move at the same speed as the steps.
- Side panels should be non-reflective. Back illuminated panels can be visually confusing.
- Contrasting nosings should be visible when both ascending and descending.
- Stationary escalators used as stairs will present difficulties for many people due to the varying and deep height of the risers.
- Lifts should be located near to escalators and clearly signed from the escalator.

> Guide dog users are often prevented from using escalators, not because of their visual impairment, but because of the need to carry their guide dogs when travelling on the escalator. This can be a difficult task for many guide dog users, and impossible for some.

Passenger conveyors (travelators)

Some people will be unable or unwilling to use passenger conveyors and an alternative form of access should always be provided.

Design requirements for passenger conveyors will be similar to those for escalators. It is critical to provide warning of the end of the journey.

- There should be good colour and luminance contrast between the moving surface and the fixed floor and good lighting in this area.
- Audible announcement and visual clues should be used to give warning of start and end of moving surface.
- There should be handrails and guarding along each side of the conveyor.
- If inclined, the gradient should be low. Steep gradients can be difficult to use when stationary and cause wheelchairs to tip up.

Figure 4.42 Where a travelator is inclined the gradient should be low to allow easy and safe use.

Platform lifts and stair lifts

Platform lifts and wheelchair stair lifts can be used to overcome changes of level where it is not possible to use a ramp or passenger lift. Stair lifts with a seat, without a wheelchair platform, are generally used in private dwellings, although they may be provided for disabled employees in the workplace.

> If a stair lift is installed, there will often be a resulting loss of a handrail to the stair. This can be difficult, especially in housing, where providing such facilities to meet the needs of one user may impinge upon the ability of another person in the house to use the stairs safely.

Platform lifts   Platform lifts have a guarded platform that travels vertically. They are used mainly to travel short distances and should be easy and safe to use by an unaccompanied wheelchair user. Platform lifts should conform to BS 6440:1999.

- Ideally lifts should be located adjacent to the stair with which they are associated.
- Approaches to the lift should be kept clear at all times.
- Clear concise instructions on lift use should be provided, including how to summon help if required.
- Lifts should be fitted with an alarm in case users get into difficulty.
- Maximum rise is 2 m in public buildings without a lift enclosure or floor penetration, or up to 4 m in private dwellings or in public buildings where there is a lift enclosure.
- Minimum clear dimensions of the platform should be 1050 mm wide and 1250 mm long.

Wheelchair stair lifts   Wheelchair stair lifts travel up the pitch line of a stair. Their use is not recommended in new buildings, though they may be useful in providing access in existing buildings where it is not possible to fit a passenger lift or a platform lift. Wheelchair stair lifts should conform to BS 5776 and in particular BS 5776:1996, Annex A. Prior to the installation of a stair lift in a public building, the Building Control Authority and the Fire Authority should be consulted and means of escape requirements met.

Assistance is usually required to use the lift and access to the stair will be limited while the lift is in use.

Wheelchair stair lifts can be installed to straight and curved flights and can continue across landings. Approach to the platform can be straight or lateral. Lateral approach will require a wider platform.

- In parked position the stair lift should not obstruct the required clear width of the stair.
- When installed in new buildings, there should be a minimum clear width of 600 mm between the folded down platform of a wheelchair stair lift and the handrail opposite.

- A means of summoning assistance should be provided. Stair lifts in public buildings should be fitted with an alarm that conforms to the requirements of ISO 9386-2.
- A fold-down seat is useful for ambulant disabled users.

Passenger lifts

The provision of suitable, user-friendly lifts will have a major impact on the accessibility of a building. Lifts are an essential amenity for many people and can provide good, easy access for everyone.

> Visual and tactile information should be used to identify the location of a lift. However, signs in and around the lift should be kept to a minimum to allow essential signs to be prominent.

The location, size and number of lifts should suit the building type and anticipated traffic. Wherever possible, there should be a minimum of two passenger lifts available for use in public buildings, to allow for one lift being out of action due to maintenance or breakdown. Where only one lift is provided for public use, consideration should be given to making available a staff lift as a temporary alternative.

- All lifts should be safe and easy to use and their location should be clearly marked.
- There should be adequate manoeuvring space outside the lift to allow wheelchair users to turn to reverse into the lift car, or to turn having reversed out. A clear space of at least 1500 mm by 1500 mm is recommended.
- A lift car 1100 mm wide by 1400 mm deep will accommodate one wheelchair user and one accompanying person. There will be insufficient space to allow the wheelchair user to turn conveniently. A wheelchair user entering forwards will need to reverse out and this should be taken into account in landing design.

The current minimum recommended lift car dimensions are 1100 mm by 1400 mm. If the lift car is smaller than this, it may still be suitable for a wheelchair user who is able to operate the control buttons without assistance. If the lift cannot be used, working practices could be altered in the short term, for example, a wheelchair user could work on the ground floor of the building instead of one of the upper floors. In the longer term, a new lift should be installed with the appropriate dimensions.

- A lift car 2000 mm wide by 1400 mm deep will accommodate any type of wheelchair, together with several other passengers. There is sufficient space for wheelchair users and people with walking aids to turn through 180°.

Where a lift is used for access between two floors only the provision of opposite doors will allow wheelchair users to enter and leave more easily.

Lift cars must meet the floor level to provide level access and egress into and out of the lift. Uneven levels can be a potential hazard, particularly for visually impaired users, people with ambulant disabilities and wheelchair users.

   The lift machinery should be checked and adjusted if necessary as part of a programmed maintenance policy.

Doors

- Lift doors should be clearly distinguishable from their surroundings using colour and luminance contrast.
- Doors should give a minimum of 800 mm clear opening width (820 mm preferred).

In certain situations where wider wheelchairs are commonly used, for example, sports wheelchairs in a sports centre, a wider lift door may be needed.

- Door control should be by photocell to ensure closing does not begin until doorway is clear. Audible warning devices announcing door opening and closing should be provided.
- Lifts should have a dwell time at landings of 5 seconds after doors have opened.

The doors to the lift often present the greatest hazard to the disabled user. The minimum dwell time to allow a person to enter the lift before the doors begin to close is 5 seconds. A dwell time of less than 5 seconds can cause people to feel anxious and worry about being caught in a closing door.

Controls and indicators

- Call and control buttons should be located between 900 mm and 1100 mm from the floor, and at least 400 mm from any return wall. Buttons should be of sufficient size, 20 mm to 30 mm diameter, for easy identification and activation, have colour and luminance contrast with their background and give tactile information, embossed, minimum 1 mm.
- The lift control buttons should not be touch sensitive.
- There should be audible and visual indication of button activation, lift arrival and direction of travel. Inside the lift there should also be audible and visual indication of floor level reached.
- Floor level indicators should be positioned so that they can be seen when the lift is full.
- Lift control panels should be located on both side walls of lift car, within reach of a wheelchair user who is facing the rear wall of the lift.

Figure 4.43 Clear signage, with information in Braille where appropriate, must be provided in a lift to inform users of the method of obtaining help and the procedures that will be adopted in bringing assistance to them.

Figure 4.44 Lift buttons located horizontally can be easier to reach.

- Vertical strips of lighting on either side of the control panel creates glare and should be avoided.
- If there is a bank of lifts, there should be a clear indication of lift arriving and time for all users to approach and enter the lift.
- Outside the lift there should be tactile and visual floor level indicators visible to users when the lift door opens and adjacent to the lift call buttons.
- Buttons essential for lift operation should be positioned separately from those rarely used.
- Alarms should have tactile information and adequate colour and luminance contrast. Alarms should have visual and audible indication to acknowledge request for assistance.
- Any emergency system that relies upon audible communication should be supported with text information, such as a textphone or written information explaining emergency procedures.
- Emergency telephones should be fitted with induction couplers to allow use by hearing aid users.

Fittings

- A support rail, 35–50 mm diameter, located 900 mm above floor level, should be provided to the rear and side walls to assist navigation for visually impaired people and as a physical support for older people or those who experience difficulties with balance.
- A mirror fitted to the rear wall, above handrail level, will help wheelchair users who have to reverse out and will allow reading of floor level indicators located above the lift doors. A mirror covering the whole wall should not be used, as it may be confusing for people with visual impairments.
- The floor surface should be firm to allow easy wheelchair manoeuvre and slip resistant.

Slip-resistant floor surfaces in lifts are of particular benefit to wheelchair users, people who use walking aids such as sticks or crutches, and older people.

- Visually and acoustically reflective wall surfaces should be avoided as they can cause problems for people with sensory impairments.
- Fold-down seats could be provided in larger lifts.
- Lighting within the lift car should be a minimum of 100 lux (200 lux preferred). Lighting should be even and free of glare.

> Where it is impossible to provide lift access in an existing building it may be necessary to relocate or duplicate essential services and facilities to ensure that they are accessible to everyone.

Evacuation lifts   Passenger lifts that are to be used for emergency evacuation should have an independent power supply and conform to the relevant recommendations of BS 5588.

Facilities

WCs

For many building users, the accessibility of an environment will be determined by the ability to locate and use the WC facilities. All WC facilities should be located and designed to allow easy use. Disabled people should be able to find and use WC facilities as easily as any other building users.

Accessible WCs may be designed to meet the space requirements of wheelchair users, but many other people use them. People with assistance or guide dogs, carers with children and pushchairs, people with luggage or shopping bags all express a preference for using an accessible compartment because of the additional space. Building managers should ensure that this use does not prevent disabled people from having access to a WC. Designers should consider providing larger compartments in the general provision of WCs for use by the above groups.

General provision

- Doors and lobbies should be designed to allow easy access to all WCs.
- Fittings, including sanitary ware, grab rails, hand driers, coat hooks and door furniture should contrast with their background.
- Cubicle doors should be clearly visible with colour and luminance contrast.
- Door locks, flushing controls, toilet paper dispensers, taps and all other fittings should be able to be used by people with limited manual dexterity.
- Floor finishes should be slip resistant and walls and floors should be non-shiny.
- General lighting level should be at least 100 lux.
- Cubicles, especially those with inward opening doors, should be large enough to allow comfortable use.
- In buildings used by carers and children one or more larger cubicles will allow assisted use of the WC.

WCs for ambulant disabled people   At least one compartment in each range of male and female WC accommodation should be designed for ambulant disabled use.

- The compartment should have internal dimensions of at least 800 mm by 1500 mm.
- Door should open outwards. If the door must open inwards the length of the compartment should be increased to give at least 450 mm diameter space clear of door swing.
- There should be horizontal grab rails to both side walls, 600 mm long, set at 680 mm above floor level with their centre line 650 mm from the rear wall. A 600 mm long vertical rail should be fitted to one side wall, 700 mm above floor level and 200 mm in front of the WC pan.
- Urinals for use by ambulant disabled people should be fitted with vertical handrails to each side.

Where cubicles or urinals are provided for ambulant disabled people or wheelchair users in the male and female facilities, all doors and lobbies should be designed to give easy access.

Accessible WCs The provision of accessible WC facilities within a building is extremely important; it may determine whether the building is truly accessible for a disabled person.

There is no standard for the numbers of accessible WCs to be provided within a building. The recommended number will depend upon building type and use. However, wherever WCs are provided in non-domestic buildings, there should be at least one that is wheelchair accessible.

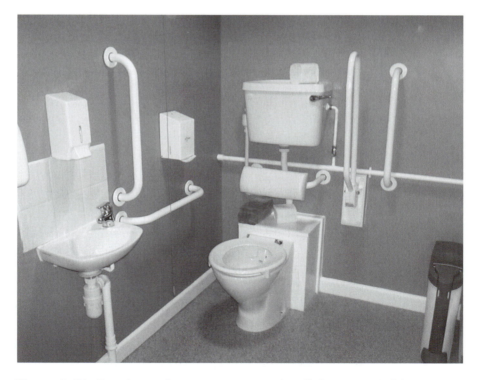

Figure 4.45 Good use of contrast ensures that fittings are visible, though position of bin reduces area available for side transfer.

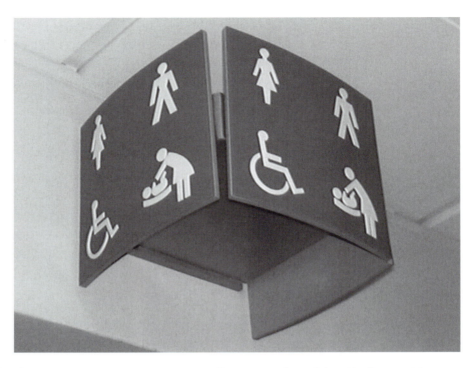

Figure 4.46 Projecting signage allows people to identify the position of a toilet from a distance, important for people who experience incontinence. This sign shows good use of colour contrast and symbols and, importantly, all the facilities are located in the same area.

Figure 4.47 A clear sign with good use of colour, symbols and text. The 'RH' identifies to a disabled person that this is a right-hand transfer toilet. This is very useful information, easily provided.

- Accessible WCs should be located close to other WC accommodation and, for visitor use, close to any entrance or reception area.
- In working environments the horizontal travel distance to an accessible WC should not exceed 100 m if on one floor or 40 m where the journey includes lift travel to another floor.
- Wheelchair users should not have to travel more than one storey to reach a suitable WC.
- When more than one accessible WC is provided, the layout should be handed. A tactile pictogram should indicate the handing.
- A unisex accessible WC, accessed from a circulation area, will allow the user to be accompanied by a person of a different gender. Where an accessible compartment is provided within the standard male and female provision, this should be in addition to a unisex WC.

> Where space is limited, for example in small business premises, a single accessible WC could be provided instead of separate sex facilities.

- The provision of accessories available in accessible WCs should be the same as in other WC accommodation.

Transfer   Space for transfer from the wheelchair is required in front of and beside the WC. It is important that boxing-in of pipes or placing of loose fittings, such as bins, does not compromise this space.

- Where boxing-in around the cistern continues across the transfer space, the WC pan should project 750 mm to allow sufficient depth of space for side transfer.

People transfer from their wheelchair from either the left- or the right-hand side, or from the front. A person that transfers from the left may be unable to use a WC that requires transfer from the right, and vice versa.

If a WC is to be provided in an employment situation, it is preferable to identify the needs of the disabled employee prior to carrying out the work, or to identify a location and have the money available to pay for it when needed. Putting in a WC when the needs of a potential employee are unknown may well result in an unsatisfactory solution – and a resultant waste of money.

Service providers should be aware that if only one accessible WC is provided within a building some disabled people might not be able to use it. Service providers may have to consider whether they have done all they reasonably can in determining the level and appropriateness of the provision available.

Layout and dimensions

- A compartment with the standard corner WC layout should have internal dimensions of at least 1500 mm by 2200 mm. The WC is located on the rear wall, diagonally opposite the door, its centre line 500 mm from the side of the compartment and the top surface of the WC seat 480 mm above floor level.

Accessible WC internal dimensions of 1500 mm by 2200 mm are minimum dimensions that, in many cases, will be too small to allow easy access for wheelchair users if equipment and facilities, such as radiators, feminine hygiene machines, disposal bins, etc., are present in the compartment.

- Where the standard layout is handed the door should be handed accordingly.

Figure 4.48 The RADAR key scheme is used to prevent the use of accessible WC facilities by non-disabled people.

Unfortunately, whilst the provision of RADAR key locks is common, not all disabled people are RADAR key holders. If this facility is provided, procedures must be put in place to ensure that disabled people who are not key holders, and who may wish to use a facility, can still do so. Whilst having a key available may be a management practice, adequate information on how to get the key is essential. Long travel distances or procedures that require a lot of time to be expended in gaining access to the WC are not acceptable.

- It is preferable that the door to the WC opens outwards to allow access to be gained if the person using the WC falls against the door. If the door must open inwards, minimum manoeuvring space (700 mm by 1100 mm) must exist within the WC and suitable hinges, which allow access from the outside in an emergency, should be fitted.
- The peninsular WC layout allows transfer to the WC pan from both sides, supported by drop-down grab rails. It is

recommended in BS 8300:2001 that this layout is only used where assistance is available and should not be provided as a substitute for two separate unisex WCs with handed layouts, but as an additional facility. Minimum dimensions are 2200 mm deep by 2400 mm wide.

Fittings and finishes

- The location of fittings such as washbasin, soap dispenser, hand drier must allow safe use from a position seated on the WC.
- The washbasin should be fixed on the side wall of the compartment at a distance of between 140 mm and 160 mm from the front of the WC pan with a rim height of 720 mm to 740 mm. There should be no pedestal.
- Horizontal grab rails around the WC should be set at 680 mm above the floor and provided as follows:

  - a drop down rail to the open side of the WC, its centre line 320 mm from the centre of the WC and extending 100 mm to 150 mm beyond the front of the WC;
  - a 600 mm fixed rail on the side wall, its centre line 500 mm from the rear wall, with a 50 mm to 60 mm clearance between the rail and the wall; and
  - a fixed rail located behind the WC with a padded backrest when the cistern is in a duct or at high level.

- Vertical grab rails, at least 600 mm in length, should be set with their midpoint 1100 mm above floor level and provided as follows:

  - to the open side of the WC, its centre line 470 mm from the centre of the WC; and
  - to both sides of the washbasin, 600 mm to 700 mm apart.

- Grab rails should be easy to grip, even when wet, and securely fixed to the wall. A diameter of 32 mm to 35 mm and textured plastic coating are recommended.
- Drop-down rails should be easy to operate, and have a vertical support set back from the front edge of the rail by at least half its length.

Figure 4.49 The use of large floor-to-ceiling mirrors can cause confusion and should be avoided.

- A small shelf for colostomy bags should be provided close to the WC at 950 mm above floor level.
- A lever-type flush with a spatula style handle should be used, located on the transfer side of the cistern.
- Paper dispensers and other fittings should be set at a height of 800 mm to 1000 mm to their lower edge.
- Where a heated drier is provided, it should not be one that is operated by movement, it should have manual, push-button operation.
- A paper towel dispenser should always be provided for people who cannot use a heated drier.
- Fixtures and fittings, including all sanitary ware and grab rails, should contrast with their background and be clearly visible.
- A horizontal rail should be fixed on the inside face of an outward opening cubicle door to wheelchair users to pull the door closed as they enter the compartment.
- Floor finishes should be slip resistant and walls and floors should be non-shiny as this can be visually confusing.

- Seats should be designed and fixed to allow for heavy-duty use. Gap-front seats should not be used.
- Mixer taps designed for use by people with poor grip are recommended or, where appropriate, automatic water supply can be provided.
- Feminine hygiene machines and incontinence pad dispensers should be provided adjacent to the WC. A sealed container for used incontinence pads and other disposable items should be provided, fixed securely to a wall, in a manner that does not affect the manoeuvring area within the compartment. A disposal bin for other general items should also be provided.
- Two clothes hooks should be provided at 1400 mm and 1050 mm above finished floor level.
- A small general-purpose shelf is useful, set at 700 mm height, away from the wheelchair manoeuvring area.
- A mirror should be provided above the wash hand basin that is suitable for use from seated and standing positions. A shaver point should be provided at the side of the mirror between 800 mm and 1000 mm above finished floor level.
- A full-length changing bench will be useful to some disabled people and should be incorporated where possible and/or appropriate. The bench should be located outside the minimum dimensions of the cubicle.
- Facilities, such as vending machines or razor sockets, available in the standard WC provision, should also be available in the accessible WC. If this is not possible, perhaps due to lack of space, they should be provided in a part of the standard WC facility that is accessible to everyone.

Emergency assistance alarm

- Accessible WCs should be fitted with an emergency assistance red-coloured pull cord, or other suitable device, capable of being operated from the WC or from the floor. The pull cord should be provided with two red bangles of 50 mm diameter (or open triangles with 50 mm sides), one positioned between 800 mm and 1000 mm and the other at 100 mm above the finished floor level. Alarms should be audible as well as visual and connected to a point of assistance.

- Within the accessible WC there must be both visual and audible feedback to show that the alarm has been activated once the cord has been pulled. This is also useful if the alarm has been activated by accident, perhaps confused with a pull cord provided for the light. A reset button must be provided which can be reached when seated either on the WC or in a wheelchair.
- The alarm may be connected directly to a staffed position, for example a reception desk or security office, or may activate an alarm indicator outside the WC compartment.

A clear written procedure must be in place for when the alarm is activated. Staff must be fully conversant with the procedure and, if the alarm indicator is sited adjacent to the accessible WC, a notice clearly indicating what should be done if the alarm is activated should be placed on or near the compartment door. Simply opening the door from the outside without properly checking that the alarm has not been activated inadvertently could cause extreme embarrassment to the user and is not an acceptable practice.

Accessible urinals   Some wheelchair users will use a urinal from a seated position or pull themselves up to use it where grab rails are provided.

- A wheelchair accessible urinal should have a rim height of 380 mm maximum and 200 mm height of clear space below. The rim should project at least 360 mm from the wall face.
- A level space 900 mm wide by 1350 mm long is needed for a wheelchair user to pull up to the urinal.
- Vertical grab rails 600 mm long should be provided on each side, 900 mm apart, height to centre line 1100 mm. A horizontal rail can be provided above the urinal.

Figure 4.50 Not all disabled people use accessible toilets; some will use the standard toilet accommodation. The need for good design, colour contrast and appropriate fittings is also important here. Children should also be considered.

Colour contrasting the urinal with the wall will greatly assist identification of the horizontal and vertical position of the urinal. Colour contrast at floor level is also useful. This vital information can be provided at no extra cost during a new build or in a redecoration.

Modesty panels between urinals can also help men with visual impairments orientate themselves and, if provided in addition to colour contrast, can give good, accessible provision.

Baby-changing facilities

- Baby-changing facilities should not be incorporated in wheelchair accessible WCs where there is only one accessible compartment. The compartment should be kept free for use by

disabled people with separate baby-changing facilities provided in a location that is accessible to male and female carers.
- Where a baby-changing bench is provided in an accessible compartment, it should not compromise space requirements.

Fold-down baby-changing benches are often left in the open position and can prevent wheelchair users accessing a WC. Careful management of such facilities, especially in WCs provided for public use, is essential.

Facilities for working dogs

If people using working dogs are employed, or visiting, a building an external area should be provided for the dogs to 'spend'. An area at least 2 m by 3 m should be identified with a sign such as 'For the use of working dogs only' and have a surface such as grass or bark chippings, a bin for dog waste, a supply of plastic bags and, where possible, a wash basin. There should also be procedures in place for cleaning the area.

Changing areas

The design of changing and shower facilities should allow independent use and a choice of communal or private facilities. Many disabled people will be happy to use communal changing facilities, but some may prefer the privacy of a self-contained cubicle.

- A changing area should have at least one accessible WC and choice of communal or private shower area.
- It is useful to include one or more unisex, self-contained, wheelchair-accessible changing rooms with integral shower, WC and emergency assistance alarm.
- Lockers should be accessible to wheelchair users, with controls no higher than 1150 mm above floor level and sufficient space for approach and use. Some lockers of at least 1200 mm high should be available to take walking aids, etc.

Further guidance is given in BS 8300:2001.

Bathrooms and shower rooms

Bathrooms suitable for use by disabled people are required in non-domestic buildings such as hotels, halls of residence and some sports buildings. This section deals with accessible bathrooms intended for independent use only. Bathrooms in hospitals and re-sidential homes will have specific requirements for assisted use. See BS 8300:2001 and other relevant guidance for further information.

To allow wheelchair users and ambulant disabled people to use an accessible bathroom or shower room independently the relationship of the fittings and the space available for manoeuvring is critical. As people have differing needs a choice of bathroom layout should be provided where possible.

- All accessible bathrooms should contain a WC.
- Where only one accessible bedroom with an en-suite bathroom is provided, the bathroom should contain a shower rather than a bath, as some disabled people will be unable to use a bath.
- Where more than one accessible bedroom with en-suite facilities is provided, there should be a choice of bath or shower and a choice of left- or right-handed transfer to the WC.
- The minimum internal dimensions for a bathroom incorporating a corner WC are 2700 mm by 2500 mm and for a shower room with a corner WC are 2400 mm by 2500 mm.
- Baths should have a flat slip-resistant base, a transfer seat, a rim height of 480 mm and a horizontal or angled support rail.
- Showers should be fitted with a tip-up seat, grab rails, and a shower curtain enclosing the seat and a shelf for toiletries.
- Flooring in bathrooms and shower rooms should be slip resistant when both dry and wet.
- Hot water temperature for bath, basin and shower should not exceed 41°C.
- An emergency assistance alarm should be provided with a pull cord reachable from the bath or shower and adjacent floor area.
- All other fittings and finishes should be as recommended for accessible WCs.

Further guidance is given in BS 8300:2001.

Bedrooms

Bedrooms in non-domestic buildings such as hotels, motels or student accommodation should allow easy use for as wide a range of people as possible. Wheelchair users have certain requirements for space and accessible sanitary accommodation and so a proportion of rooms should accommodate such use. Wheelchair-accessible rooms should be provided in a choice of locations and on accessible routes that lead to all the other facilities within the building.

- Accessible bedrooms will require sufficient space for manoeuvre around the room and transfer to one side of the bed.
- Wheelchair users should be able to use all the facilities in the room, including balconies, and should be able to gain access to and use en-suite sanitary accommodation.
- Wheelchair users should be able to visit companions in other rooms.
- It is desirable for some accessible rooms to have a connecting door to an adjacent room for a companion or assistant.
- Accessible bedrooms should be fitted with an emergency assistance alarm operated by a pull cord that can be reached from the bed and from an adjacent floor area.

Storage facilities

Disabled people may use storage facilities such as cupboards, shelves and lockers when at work or visiting buildings. Access to storage units, height of units and shelves, knee space and ironmongery should all be designed to allow easy use, with specific facilities provided where appropriate.

- Wherever storage facilities are provided for the general public there should be at least one fully accessible facility for disabled people.
- Wherever possible some knee spaces should be provided to allow a choice of frontal or sideways use from a sitting position.
- The corridor width between banks of storage units should be 1200 mm if some knee spaces are provided and 1400 mm if no knee spaces are provided.

- It is useful to provide some seating at storage facilities for use by ambulant disabled people.
- Ironmongery should be easy to grip and manipulate and contrast in colour and luminance with its background.
- Where storage facilities are provided for a particular person, such as in a place of employment, they should be designed to suit the individual's needs.

  BS 8300:2001 contains information on reach ranges for wheel-chair users and ambulant disabled people.

Refreshment areas

Restaurants and cafeterias should be designed to be accessible in terms of both spatial design and management. Staff training is critical in areas such as these where assistance may be required to overcome design limitations.

It may be difficult to find vending machines with all the correct requirements, but the employer or service provider should ensure that the machines they are providing are the most accessible that are available, or that they develop management procedures to assist those people who may experience difficulties when using them.

There may be a temptation to remove facilities so that the provider is not discriminating against disabled people – because no one, whether disabled or non-disabled, is being provided with the service. That should not be necessary; in the vast majority of cases, careful thought and good management can overcome, or at least address, any temporary shortfall in the accessibility of products and facilities.

- Split-level areas should be linked by ramps.
- Where there are fixed tables there should be sufficient space for wheelchair users to circulate and have a choice of seating locations.

- Fixed seats will be difficult for many disabled people to use. Loose seating is preferable and can allow space for wheelchair users to pull up at tables. At least some seats should have armrests.
- Vending machines should be designed and located to allow approach and use by wheelchair users.
- Self-service counters should be continuous. Where trays need to be moved from one counter to another, assistance should be available.
- Self-service cabinets and shelves should be located within reach of wheelchair users wherever possible.

People who are deaf or hard of hearing often experience problems at cashier desks because they may be unable to follow the prices being added up and unable to hear how much is required to be paid. Simply orientating the display on the cash register so that the user can see the details of the transaction will greatly assist.

Induction loops are often of little benefit at cash register points because of the interference caused by the electrical equipment being used there.

Good lighting and staff training are essential at such points.

Counters and service desks

Counters and service desks should be accessible to wheelchair users as staff and customers.

- There should be sufficient space for wheelchair users to approach and turn towards the desk.
- The knee space and height of desk should allow close approach where this is required. Seating should be provided at low height counters.
- If there is a glazed screen it should be able to be opened to allow for direct communication. If it cannot be opened, or if

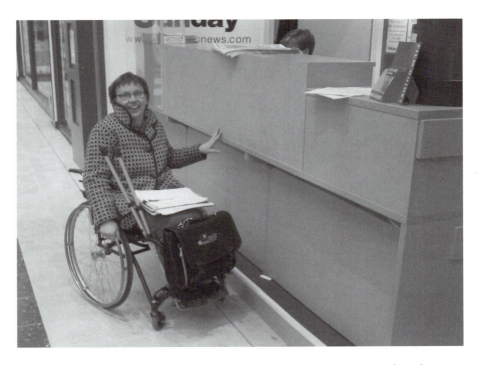

Figure 4.51 Counters should allow communication at seated and standing heights. A low counter or low section will allow wheelchair users to communicate with staff or customers. Seats should be provided to both staff and customer side.

there is background noise, an induction loop should be provided for the benefit of hearing aid users.
- The faces of staff employed at the counter should be well lit to allow for lip reading (see also *Reception desks*).

Assembly areas

All assembly areas should allow access and use by disabled people as members of an audience, participants and members of staff. Disabled people should have access to the full range of seating options and be able to sit alongside disabled or non-disabled companions.

Guidance on seating layouts is given in *Reception areas*. In some circumstances space for lying down may be required for people who are unable to sit or stand for long periods of time.

Lecture rooms, conference facilities, meeting rooms

- A mixture of fixed and removable seating will allow for variation in the number of wheelchair users accommodated.
- Provision should be made for wheelchair users as members of the audience and speakers.
- Any raised area or podium should be wheelchair accessible.
- Lecterns and other fittings and controls should allow use at a variety of heights.
- Good sight lines and lighting are particularly important to allow for lip reading and the interpretation of sign language.
- A hearing enhancement system should be provided. These systems are described in *Communication and wayfinding*.

Lecture rooms with raked floors

- Wheelchair spaces should be provided in a choice of locations and should be provided with a handrail and crash bar at any change of level.
- Access routes should be provided with handrails.

Auditoria, stadia and spectator seating

- Disabled people should have access to the full range of seating locations and be able to sit alongside a disabled or non-disabled companion.
- Routes should be accessible and of sufficient width to allow wheelchair users and ambulant disabled people to circulate.
- Handrails should always be provided to stepped and ramped routes.
- Numbers of spaces for wheelchair users should be in accordance with the guidance given in BS 8300:2001 and Part M of the Building Regulations. In addition, there should be transfer seats provided for wheelchair users who wish to transfer from their wheelchair to a seat for the duration of the performance.
- Space should be available for working dogs adjacent to seating and clear of circulation routes.

- Wheelchair seating areas in sports stadia should be designed to allow wheelchair users to see the event even when people in front stand up.
- All wheelchair seating locations should have access to an accessible unisex WC.
- A hearing enhancement system should be provided.
- Emergency egress procedures should take account of the needs of all people who need assistance, whether they are seated in designated areas or not.

Sports venues   Facilities should be provided at sports venues to allow disabled people to participate and compete in all the available sports. Guidance on circulation and provision of facilities should be followed.

- Sports wheelchairs are often wider and longer than standard wheelchairs and so additional circulation space is required. Door widths and lift sizes should also be increased.
- There should be level access from changing facilities to sports areas and swimming pools.
- Access into swimming pools should be provided with a choice of methods, such as ramped access or hoist.
- Hearing enhancement should be provided in fitness and exercise areas where instruction may be given to participants.

A radio receiver system is often used in sports stadia to provide an audio description service for spectators.

Controls and equipment

Ease of use   All building users should be able to operate the controls and equipment provided for their use. Ease of operation depends upon these items:

- being within reach;
- not requiring excessive strength or dexterity;

- having clear instructions for use that are visible, and where appropriate audible and/or tactile.

Coin and card operated devices   Coin and card operated devices such as vending machines and automatic teller machines (ATMs) should be designed and located to allow easy independent use.

- The height of controls should allow access from a seated or standing position, with knee space provided where possible to allow wheelchair users to pull up facing the front of the device.
- Controls should allow use by people with limited manual dexterity and sensory impairments.
- Display screens should be visible and shaded to prevent glare and reflection.
- Clear operating instructions should be provided.

Building services – sockets, switches and controls
Building service controls should be designed and located to be able to be seen and used easily. Ideally, controls such as switches and sockets should be located consistently throughout a building at a height within the reach range of all users.

- Controls should not require the use of both hands simultaneously.
- Controls should contrast in colour and luminance with their background and have embossed tactile information where appropriate.
- The use of large touch plates and rocker switches will benefit people with visual impairments and limited manual dexterity.

Light switches will be easier to locate when entering a room if they are aligned horizontally with the door handle.

Alarms   The design of alarm systems within buildings should take account of the needs of all building users. Alarm activation should be able to be carried out by wheelchair users. People with hearing impairments should be aware that alarms have been activated or announcements made.

- Where audible alarms are used they should be supplemented with visual systems if located in areas where people with hearing impairments are likely to be alone.
- Personal vibrating pagers can also be used to supplement an alarm system.
- Alarm call points should be located within reach of seated and standing users and designed to allow easy use.

Vibrating pagers can be useful and cost effective, especially if required by only a few people using a building. However, there are management implications that need to be considered, such as ongoing testing and the replacement or recharging of batteries.

In addition, there will be issues of how and when the pager will be used. For example, pagers that attach to a belt are unlikely to be easily usable by everyone. Pagers need to have direct contact with the person using them to be effective; those placed in handbags, jacket pockets or in desk drawers are very unlikely to be of any use.

Supplying pagers is not sufficient – providing ones appropriate to the needs of the user is essential, as is good management of the whole process to make sure it works effectively.

- Where emergency assistance alarms are located in WCs, bathrooms, changing rooms or hotel bedrooms, there should be visible and audible feedback to indicate the alarm has been activated.
- Emergency assistance alarm indicators should be located where they will be seen and heard by those able to give assistance.

Windows and window controls   The location of windows and window controls and the ease of use of controls will affect the ability of people to use windows effectively. An appropriate sill height and controls that are easily operated and located within reach will allow wheelchair users, ambulant disabled people and others to use windows independently.

- Window furniture should be visible and located within reach.
- Controls should be able to be used without excessive strength.
- Powered systems can be used to eliminate the need for manual opening and closing.

In many window designs, a fanlight is incorporated to allow a small amount of ventilation. When it is raining, using a fanlight may be essential as open casements may allow water to penetrate into the room.

The opening mechanism to any fanlight should be placed within easy reach of all users, or another suitable method of opening provided.

Window controls that require the user to climb to operate them, for example when fitted furniture is located in front of the window, must be avoided. Remote opening either by a manual or powered means should be provided.

Communication and wayfinding

Building design can help or hinder communication. The use of lighting, colour and luminance contrast, signs, hearing enhancement systems and the provision of information in a variety of formats will all affect the ability of people to use a building independently. It is not just people with sensory impairments who benefit from good design in these areas; it is everyone.

Access to information

It is important that people have access to information, including signs, leaflets and verbal information.

Most information in buildings is given visually, but the use of other senses should be considered and acoustic, tactile and olfactory factors incorporated into building design. A sound such as a water feature may be a useful landmark in a large office building. A person with a visual impairment might use the sound of lift doors or an activity taking place at a counter to assist in locating the service.

- A high level of background noise should be avoided where possible as it can cut out useful noises that may help with navigation.

Tactile signs or maps will be useful to some people, if located where they can be found and touched. See sections on signs and tactile surfaces for further information.

Olfaction is a valuable source of information, currently under utilised inside buildings, that can aid navigation if used appropriately.

- Cedar and other woods used in building have a distinctive smell and fragrant plants can also be used to provide pleasure and orientation clues for people with visual impairments.
- Air movement and temperature can also give useful clues to help with orientation.

Environments that encourage feelings of comfort and well-being are important. The use of a variety of sensory information can make environments more pleasurable for everyone, as well as providing useful information to people with visual impairments.

Visible reinforcement

- Answerphone systems, bells and buzzers should have visible indicators to confirm their operation.

- Public address systems should also give visible information using, for example, an LED screen to supplement the audible message.
- Alarm systems should also take into account the needs of people with hearing impairments and should be backed up by visible confirmation.

Written information

- Leaflets should use clear print and information should be provided in other formats where necessary. This could include tactile maps, information in Braille or on audiotape.

It can be very useful to visitors to provide information about the accessibility of a building in advance of a visit. An access leaflet with information on public transport, parking arrangements and facilities available within the building will allow disabled visitors to plan their visit and be aware in advance of any potential problem areas. This information itself should be available in a variety of formats.

Not all literature has to be printed in accessible style and type – but accessible formats, including large print and audiotape, must be available if requested. Information in Braille may not always be necessary because people who read Braille may also be able to get the information they need using an audiotape. However, if required, it should be provided.

Wayfinding

Building layout affects wayfinding. A logical straightforward layout will allow people to navigate more easily, remember routes and can assist means of escape. The wayfinding process is dependent upon receiving and processing information; the more quickly and accurately people are able to do this, the more rapidly the journey will be completed.

Figure 4.52   Identifying the presence of facilities with carefully designed lighting can greatly enhance the information available to disabled people – even when signage is not used.

When living in and moving through environments, most people use the five senses of sight, hearing, touch, smell and taste to gain information. It has been estimated that 70–75% of information people receive is through vision and that 10–15% comes through hearing. The remainder is gathered through the remaining senses.

For people who have significant sight loss, the 70–75% of visual information will be reduced, and completely eliminated in the case of those who are totally blind. However, the problem is not restricted to people with visual impairments. Anyone with more than minor hearing loss will need to obtain information through visual clues, so that they can avoid having to ask for directions, knowing that they would have difficulty in hearing the reply.

- Signs, colour and contrasting tactile information can all be used to help with navigation.
- Colour can be used to identify routes or differentiate between areas or storeys of buildings.

- Colour and luminance contrast can help distinguish surfaces and features.
- Tactile information can be given by floor surfaces, embossing on signs or the provision of maps and models that enable visitors with visual impairments to understand a building shape and layout.
- Audio guides can be provided, as in some visitor attractions.

Tactile surfaces   Tactile paving can impart useful information when used correctly. A number of different profiles have been developed, each of which has a specific meaning. Care should be taken to install the appropriate texture, colour and hardness.

- The most commonly seen pattern is the modified blister, which is used externally at dropped kerbs or raised road surfaces to indicate the edge of the pavement. Different colours indicate the presence of a controlled or uncontrolled crossing.

Figure 4.53 The incorrect use of surface finish can completely negate the effectiveness of a sign, especially in certain lighting situations. Non-reflective surface finishes for signs are essential.

- The hazard warning pattern, corduroy, a half-rod shaped profile, is used at the top and bottom of stairs.
- A flat-topped ribbed profile denotes a guidance path; this can be used in large open areas such as pedestrian precincts.
- Non-specific changes of profile and texture in internal and external surfaces can be used to give general information on routes and to guide people clear of hazards in circulation areas.
- Tactile features can also be used on signs, maps, controls and handrails.

Human interaction   The attitude adopted by those managing an environment, often first encountered at a car park entrance, reception or 'help desk', can have a major impact on the accessibility of an environment for all users, but especially users with disabilities.

Staff should be trained to understand the needs of particular user groups, how to communicate effectively and how to offer assistance when required. This is often a very cost-effective way of improving dramatically the way in which disabled people can safely, effectively and independently use an environment.

- Management responses to requests for assistance to use an external lift, approach or enter a building, should be devised, practised by staff and operable at all times.
- Where a control device contains an audible communication system, to announce arrival or seek assistance, the needs of people who are deaf or hard of hearing must be considered. Clear information should be given at the control device and a management response available at all times.
- Staff should be trained in dealing with the communication needs of people who are deaf or hard of hearing.
- Reception staff should know of the location of facilities such as accessible WCs.
- Key staff should understand the needs of people with visual impairments and undergo basic mobility training to assist them in moving around an environment safely if required.
- Key staff should be trained in clear speaking and, if possible, basic sign language, to assist people who are deaf or hard of hearing.

- Staff should be fully trained in the use of any mobility equipment, such as platform lifts. Records should be kept of who has been trained and assistance should be available at all times.
- Visitors should be informed when there is assistance available.
- Staff should be aware of the potential problems of certain actions, such as the use of 'home-made' temporary signs that do not follow good practice design guidance or the positioning of obstacles in circulation spaces.
- In existing buildings where access barriers may be difficult to eliminate completely, staff input will be critical.

The importance of good staff training and the ability of staff to be able to recognise that a difficulty may exist, and to be able to move swiftly, and appropriately, to address the issue cannot be overemphasised.

Not all buildings are accessible. Disabled people know this and most acknowledge that not everything can be perfect. However, what is important to them is to be able to experience an equal opportunity to that experienced by non-disabled people. The interface between staff and disabled and non-disabled users of a service is critical in how imperfections in an environment are dealt with, and whether the obstacle met is acceptable or unacceptable, surmountable or insurmountable.

Signs If a building is designed in a logical and simple manner, the need for signs is minimised. The design of the external environment should lead people to the entrance and internally there should be clear, logical routes and access to services. Where signs are needed, they should be well placed, well lit and use clearly visible print.

Signs are of particular importance to people who are deaf or hard of hearing, who often will not ask for directions as they may be unable to hear the answer. Signs will also benefit older people, those with learning difficulties or short-term memory loss, who may need reassurance they have taken, and are still on, the correct route.

Figure 4.54 Signs should be clear and easy to understand. Too many signs can cause confusion.

Well-contrasted, correctly sized signs with lower case lettering and, where appropriate, tactile (embossed or Braille) information will also assist those with restricted vision. Colour coding and symbols can be useful if used consistently.

However, in reality, all users will benefit from a signage system that is well designed and properly co-ordinated.

- Signs should be simple, short and easily understood.
- Signs should be consistent, using prescribed typefaces and graphic devices.
- Capitalised lower case lettering is generally easier to read.
- The typeface used should be clearly legible and preferably sans serif.
- The number of different typefaces and font sizes on a sign should be kept to a minimum.
- The size of lettering should be appropriate to the distance from which the sign is viewed.
- The signboard should contrast with its background and the lettering should contrast with the signboard.

Figure 4.55 It is important to avoid 'information overload'. Effective, ongoing management of signage is essential. Temporary signs should be removed and, if necessary, replaced with permanent signage.

If signs are colour contrasted with their background, this can be particularly helpful to visually impaired people and also to make them more readily identifiable for people who are deaf or hard of hearing. The latter group rely heavily on signage for information, as they are often reluctant to ask directions in case they will be unable to hear or decipher the response.

- Light lettering set on a dark background is generally easier to read.
- Locating signs at eye level with easy access for close viewing will benefit everyone.
- To avoid glare signboards should be non-reflective and lighting positioned appropriately.
- The use of appropriately positioned tactile maps or plans can be of benefit to all users, but especially those with restricted

Figure 4.56 Good clear signage, though the word Disabled is unnecessary. Toilet or WC accompanied by the symbol is sufficient.

Figure 4.57 Good use of text, contrast, symbols and directional arrows – and ideally sized, bearing in mind the distance from which it will be viewed.

vision. Tactile signs should be positioned where they can be easily reached. Tactile information should be embossed between 1 mm and 1.5 mm.

- Information in Braille should be used in addition to embossed information, not instead of, and should be kept to the minimum required.

The number of people who read Braille is small compared with the number who would benefit from the appropriate use of well-designed embossed lettering, numbers or symbols. In addition the way that Braille is normally read, i.e. horizontally at table or lap position, means that providing too much Braille on a vertical sign will actually be difficult to read and, in most cases of little benefit to Braille readers. More useful would be to provide clear, well-designed embossed signs, accompanied by a limited amount of Braille to identify what features the sign is identifying.

If larger amounts of information are needed in Braille, the delivery of it, in terms of positioning and design of the signboard, needs to be carefully considered.

- Symbols can be very useful, particularly where decisions need to be taken quickly, such as in transport environments.
- It may be necessary to complement symbols by text to clearly define their meaning; for example, a wheelchair symbol can be used to signify an accessible WC, an accessible route or parking space.

Figure 4.58 Symbols on signs can be useful for many people.

Ideally any signs should incorporate a combination of lettering and symbols.

Symbols are useful for people who experience dyslexia, those with learning disabilities and people who do not speak or understand the language of the sign. For some people with learning disabilities, information may be given totally in symbols.

As a general recommendation, specialist publications should be consulted on the design and layout of signs, and specialist designers should be employed to advise on the most appropriate signage to use.

- Lighting should ensure that signs can be seen when daylight is poor.

Detailed information on signs is given in Sign Design Guide (Sign Design Society & JMU 2000).

Figure 4.59 The use of the letters WC indicates this is an accessible toilet. Using the symbol without the letters may suggest to some people that this an area related to disability, but not necessarily a toilet.

Lighting

Good lighting is important for everyone. Lighting can be used to enhance colour and luminance contrast to increase the visibility of routes and objects and make environments more easily legible and safer to use.

The design of a building or environment should allow people with visual impairments to use the vision they have as well as providing the other factors that can assist navigation. Older people and people with visual impairments generally require higher lighting levels. Good levels of lighting are also required to allow people to lip read or use sign language.

People with visual impairments are generally sensitive to glare and have longer adaptation times than people without a visual impairment. Adaptation is the process by which the visual system changes to optimize viewing in either a darker or lighter environment. In general, moving from a dark environment to a bright environment, such as may occur when walking out of a building on a bright sunny day, or in an emergency situation when the mains electrical supply fails and emergency lighting, normally at a much reduced level, is provided, will present few problems for people who do not have a visual impairment. However, for people with a visual impairment, adaptation times are greatly extended.

Designing lighting systems to meet the requirements of an inclusive environment is possible with modern technology. When the design is based on the needs of people with visual impairments, the needs of all other users will normally be met. Controlling glare and diversity or uniformity within a space has the potential to assist all users.

- General lighting should be controllable and adjustable to meet the individual user's requirements.
- Good lighting is critical where there are potential hazards such as changes of level.
- Light sources should be located to avoid glare, reflection and strong shadows.
- Lighting can be used to help identify features and objects by emphasising shape, form and texture.
- Good lighting is important to allow communication by lip reading and sign language.

Figure 4.60 Strong shadows can cause confusion and make navigation difficult.

At a reception desk, cashier point or at any point where people may be required to complete forms or read or understand instructions, an ability to be able to control the lighting available in given circumstances is critical.

Control may be by way of providing switches which can increase or lower the amount of main lighting available, additional lighting which can be switched on when needed or by the provision of a task light. However, it is important to remember that because such a provision may not be in constant use, the switch to operate it must not be placed in some inaccessible position. When needed it may be needed swiftly, and accessible switching and good staff training are essential.

- There is no definitive guidance on lighting levels for people with visual impairments as their needs vary widely. However, lighting levels of between 25% and 50% above those given in CIBSE guidance are generally recommended.

- Where possible, windows should not be positioned where they allow glare and reflection from sunlight on stairs, corridors and other areas. The position of the windows may not be able to be altered, but window blinds could be fitted or films applied to the glass where appropriate.
- Task lighting should be controllable and adjustable to meet the individual user's requirements. Illumination requirements will vary depending upon the task requirement to be undertaken and the relative brightness of the surfaces within the space being lit. Some people, because of the nature of their visual impairment, work close to a task light and so lights that generate a lot of heat should be avoided.
- Fluorescent lighting can cause a hum in hearing aids and should be used with care to minimise inconvenience to hearing aid users.

Figure 4.61 Window blinds or shading can be used to control natural lighting and prevent strong shadows and glare, which can be disorientating.

Colour and luminance contrast

Colour and luminance contrast used in decoration gives information to building users and can be of particular importance to people with visual impairments. Different colours may have a similar luminance contrast and so it is important to consider both colour and luminance, and lighting conditions.

Colour and luminance contrast will assist most people, especially those with visual impairments, at certain points. These fall into two categories, which can best be described as those that are critical surfaces, and those that are special features. Critical surfaces are large areas that, when scanned by a visually impaired person, form the impression of space, shape and proximity. Examples are ceilings, walls, doors, floors and stairs.

Special features are additional areas, smaller than critical features that need to be highlighted to allow the building to be used easily by people with visual impairments. Such features include sanitary ware, handrails and stair nosing, door handles and socket outlets, all of which should be contrasted with the background against which they will be seen.

- The size and shape of a feature is an important clue to its identity. There are few features within a building that offer the same visual image, in terms of size and shape, as a door adequately contrasted with its surroundings.
- Navigating through a building is much easier if critical surfaces are colour and luminance contrasted.
- On critical surfaces the use of highly reflective, shiny finishes can cause considerable confusion for everyone. Such finishes should be used with caution.
- Trim items, such as coving, skirting, architrave, dado rail, etc., should be decorated in colours that maintain or improve the impact of the colours used on the larger critical surfaces.
- Special features such as sanitary ware, handrails, stair nosings, door handles, socket outlets and switches should be contrasted with the background against which they will be seen.

Identifying hazards   The vast majority of people with visual impairments say that colliding with furniture and obstacles in the walking area is a frequent and disconcerting occurrence.

Obstacles that are free standing, project from walls or overhead should be kept to a minimum and adequately contrasted with the critical surface against which they will be viewed.

Colour and luminance contrast, especially when related to the identification of hazards, is equally as important for people who are deaf and hard of hearing as it is to those with restricted visual ability.

To communicate with other people when walking or moving around, people who are deaf or hard of hearing will often look at the person they are communicating with to lip read or to sign. When doing this, they will be relying on their peripheral vision where, just as with a visually impaired person, their ability to see colour and fine detail is poor. Deaf people can, and do, trip over and collide with hazards not because their vision is poor, but because their strategy for communication means they are not always looking in the direction they are going. But doesn't that happen to most people at some time or another?

Acoustic environment

The acoustic environment is critical to allow people to use their hearing capability effectively. Many people with hearing impairments will use lip reading and so visual conditions are important also.

- The acoustic environment will affect the ability to communicate, as will light levels.
- The reduction of background noise is beneficial; however, some reflected sound can assist people with visual impairments in understanding the space that they are in and to hear others approaching them.

- Lighting and visibility should be sufficient to allow lip reading.
- Hearing enhancement systems: induction loops, infrared systems, radio systems, can be useful where information is given verbally.
- Key staff should be trained in communicating and interacting with deaf and hard of hearing people. The 'hearing aware' symbol can be used where trained staff are available.

Hearing enhancement systems   Electronic hearing enhancement is used in buildings to amplify sound or to provide additional information to users. The two most common types are induction loops and infrared systems, though radio transmitters are also used where appropriate. Standard signs should always be used to inform of the presence of these systems.

Effective use of hearing enhancement systems relies on maintenance and management procedures. Potential users need to be aware that a system exists, hence the need for signs, and staff should be able to operate the system. All equipment should be regularly checked and maintained.

> It is important that equipment, such as an induction loop installation, is properly maintained and tested. To be of use it must be working when the disabled person needs to use it.

Induction loops   Induction loops transfer sound spoken into a microphone into sound for transmission to a hearing aid. Induction loops work by converting sound via the microphone into a varying magnetic field that is converted back to amplified sound by an individual's hearing aid. They are commonly used at reception desks, counters and in meeting rooms.

- The hearing aid should be set to a 'T' position to pick up the amplified sound and so it is important that the presence of the induction loop is indicated.

- Induction loops only benefit hearing aid users and so there is still a need for good communication skills and clear written instructions as appropriate.
- An induction loop can be built into a room or counter and portable induction loop systems are also available.

The provision of induction loops needs careful consideration if money is not to be wasted on inappropriate installations. In shops, induction loops positioned at tills are unlikely to be of significant value because of interference from electrical equipment such as cash registers. Simply orientating the cash display so that someone who is deaf or hard of hearing can actually see the totals may often be the better, more appropriate, solution.

However, there will be areas where the need to transfer information clearly is essential, for example at a pharmaceutical counter where important information about the taking of drugs is being given. Here, the need for an induction loop to ensure instructions are fully understood could be critical.

Infrared systems   In an infrared system, a microphone is used to collect sound, which is sent to an amplifier and coder that converts the sound into infrared light. The receivers, usually headsets, convert the light into sound information to transmit to the wearer. Infrared systems are used in theatres, lecture rooms and other controlled environments where it is possible to manage the supply and collection of receivers.

Radio systems   Radio receiver hearing enhancement systems are portable as users carry personal receivers and are often used in schools and colleges where students move between classrooms. They are also used in sports stadia to provide a commentary to spectators. Information can be transmitted on more than one channel.

Telephones   Where public telephones are provided in a building at least one should be accessible for wheelchair users. Where there is a selection of telephones with different payment methods one of each type should be available to wheelchair users.

- Where there is only one pubic telephone it can be set at a low height provided a seat (fold down if necessary) is available.
- The maximum height of the top control or slot should be no more than 1200 mm above finished floor level.
- To assist users who may need to steady themselves when using the telephone a support rail should be fixed to the wall adjacent.
- A shelf should be provided for use of a portable text phone.
- Any instructions should be provided in formats suitable for people with visual impairments.
- Public telephones should be fitted with inductive couplers and a variable volume control.
- Wherever possible, a text phone should be provided for public use.
- Any sound covers used must allow access by all users and not be hazardous to people with visual impairments.

Emergency egress

Egress must be considered alongside access. A well-designed, accessible building should allow independent egress for as many of its occupants as possible. Safe, efficient egress will depend upon a combination of management procedures and building design.

Specific evacuation strategies may need to be devised for people who need assistance, for example where lifts cannot be used for vertical escape. These strategies should take into account the building design, the known needs of people working in a building, as well as the unknown needs of visitors.

Fire engineering techniques can be used to assess the risk of fire in any part of a building, taking into account factors such as compartmentation, intelligent alarm systems, sprinklers and smoke control. This can provide the basis for an evacuation strategy for a particular building with a particular pattern of use.

It is important to provide people with the information necessary to allow them to make informed choices. Clear warnings, signs and instructions are needed to tell people where to go and what to do in an emergency. Disabled people are often well aware of personal risk management and are more likely than others to act rationally if provided with the appropriate information.

Generic evacuation plans   The needs of visitors to a building may not be known and a generic plan should be devised to deal with the evacuation of people who may need assistance. A plan of this sort should consider the needs of wheelchair users, ambulant disabled people, people with sensory impairments and others whose needs cannot be identified in advance.

Personal emergency egress plans   Personal emergency egress plans (PEEPS) should be devised for employees who have specific needs.

- The plan should take into account the building design and how it may cause difficulties in evacuation, the type of assistance required to overcome these difficulties and any other requirements specific to that employee.
- Where assistance is required, the plan should also set out how and by whom that assistance will be given.

Horizontal and vertical evacuation   Horizontal and vertical evacuation should be considered. Some people may be able to evacuate themselves horizontally to a fire-protected refuge space, on or near to escape stairs, but need assistance to facilitate vertical escape. If an evacuation lift is not available, people may need to be assisted down stairs. Evacuation chairs can be used in this situation. These are lightweight chairs that can be operated by one person, rather than the three people required to carry a wheelchair and occupant. However, evacuation chairs may not be suitable for all wheelchair users and some people may prefer to be carried in their wheelchairs.

The use of phased evacuation in larger buildings, where people are initially transferred to a separate compartment within the building and then evacuated vertically, may allow lifts to be used for vertical escape.

Refuges   A refuge is defined by BS 5588 Part 8 as an 'area that is enclosed with fire resisting construction and served directly by a safe route to a storey exit, evacuation lift or final exit, thus constituting a temporary safe place for disabled people to await assistance for their evacuation'. The BS goes on to say that refuges are relatively safe waiting areas for short periods. They are not areas where disabled people should be left indefinitely until rescued by the fire service, or until the fire is extinguished.

- Refuges should be identified and clearly sign posted.
- Size and location of refuges need to be carefully considered. It is preferable to locate refuges on escape routes, where people using the refuge can be seen.
- There should be a two-way communication system provided to allow a person waiting in the refuge to communicate with those organising the evacuation. It is preferable to provide a visual link as well as verbal as this allows for checking refuge occupancy in an emergency situation.
- Where escape is via stairs, evacuation chairs should be provided in the refuge and staff trained to operate the chairs.

Assistance   Availability of assistance must be considered. Where a disabled employee has a personal emergency egress plan the

Figure 4.62 Refuges should be clearly identified.

assistance need is known in advance and can be planned for. The needs of visitors will not always be known and strategies must be developed to accommodate their evacuation needs.

- Where a horizontal evacuation strategy includes the use of refuges there should be planned procedures for assisted vertical evacuation.
- It is the responsibility of the building management, not the fire brigade, to evacuate people.
- Management should carry out staff training, regular reviews of plans and organise regular practices.
- There should be regular checking of escape routes and fire alarm and fire-fighting equipment.

Evacuation lifts Multi-storey buildings should be provided with at least one evacuation lift. This is a lift with an independent power supply and control and located in a fire-protected shaft that can be used for emergency evacuation.

Fire alarm systems and equipment The fire alarm system in a building should be suitable for everyone. Issues to be considered should include the type and location of manual call points, whether an automatic detection system is required and the addition of visual alarms to an audible alarm system.

- Manual call points should be located where they can be reached by everyone, including wheelchair users, and consideration should be given to providing alternative types of call mechanisms, such as pull cords.
- Audible alarm systems should be supplemented by visual means of warning for people with hearing impairments. The use of individual vibrating warning devices may also be appropriate.
- Fire extinguishers, fire alarm call devices, hoses, blankets and other equipment should be positioned to allow everyone to reach, take down and use.
- There should be clear instructions on emergency procedures and clear signs.

Figure 4.63 Fire exit signs should be located where they are clearly visible and care taken to ensure lighting does not cause reflections that obscure the information.

Emergency lighting   There is a need for emergency lighting to be provided in many buildings and the usual approach is to adopt the requirements of BS 5266. This is generally concerned with overhead lighting. Where visually impaired people are likely to use the escape route, more than the minimum 0.2 lux recommendation of BS 5266 should be provided. It is recommended that visually impaired people be provided with a minimum of 3 lux along the escape route.

In addition, there are also a variety of wayguidance systems available, which may offer an important contribution to escape efficiency in situations where the evacuation of large numbers of people may be required, for example, from buildings such as auditoria, theatres or cinemas.

Unpowered photoluminescent wayguidance systems are not currently recommended for use in areas where visually impaired people have access. In these systems, a photoluminescent material, perhaps in the shape of strips or blocks (notices, etc.), absorbs light from the main lighting system when in general use and emits it after the main lighting has gone out. The amount of light provided by the system has been found to be too low to be identified by visually impaired people.

Powered wayguidance systems are generally systems that are mounted at low level and provide a strip of continuous light or a

Figure 4.64 The use of powered wayguidance systems will not be appropriate in all buildings in all situations. However, in buildings where there may be a need to evacuate a lot of people (disabled and non-disabled), such as cinemas, auditoria, conference venues, etc., the selective use of such systems can assist considerably in improving the evacuation speed of all users.

strip of closely spaced individual light sources, e.g. light emitting diodes (LEDs). Certain powered wayguidance systems are preferred by visually impaired people to overhead systems that produce a minimum of 0.7 lux at floor level.

Although powered wayguidance systems are relatively new, they are becoming more available.

Final exit   Exit routes and exit doors should be accessible to all disabled people, including wheelchair users. Steps at final exits must be avoided if possible. Thresholds should be level and doors should be wide enough to allow a disabled person to pass through

Exit arrangements are often overlooked in access audits or not considered in sufficient detail. Where there is no proper assisted escape procedure and it is necessary for a disabled person to be physically lifted to overcome even one step at the exit, there could be serious ramifications for the managers of the building, not to mention the disabled person.

For example, lifting a wheelchair with a disabled person in it may well subject employees to an action which is contrary to that permitted under other relevant legislation concerning health and safety or lifting. Staff may well seek compensation if a management practice such as this causes injury to them. The person using the wheelchair may be injured in the lifting process and may also seek compensation.

Figure 4.65 Assisting disabled people out of the building using a fire exit with external steps presents management issues that simply cannot be addressed by an *ad hoc* escape policy.

to an equal standard as when entering the building. Wherever there are steps, internally or externally, a refuge should be provided and an assisted escape procedure devised, regularly reviewed and updated.

Appendix A
General acceptability criteria

Access audit checklists

The following checklists have been prepared as an example of easy reference guides that can be used when undertaking access audits.

The questions identified here are the more general issues of the type that should always be considered in an audit. There may well be others that also need to be considered in specific situations.

Checklists such as the ones described here should only be used as an aide mémoire, a method of recording information and a way of highlighting issues. They should never be used as the sole method of reporting on access issues.

How to use the checklists

When using the checklist, circle the response in the ✓ column for yes or in the ✗ column for no.

If the question contains the words 'if any' and none is provided – do not respond to the question.

Only respond to one of the questions posed in those which offer an OR response.

Respond to both questions for those which offer an AND response.

Sometime you are just asked to circle the ✗ if something exists. That is because there is definitely an access issue that needs to be considered and there is further information elsewhere in the manual or, in some cases, BS 8300 or Part M of the Building Regulations.

If one of the items described in the checklist achieves a ✗, then issues will need to be addressed and, probably, some work needs to be carried out. This may be physical alterations or changes in terms of practices, policies or management procedures.

affl = above finished floor level.

Accessible car parking checklist

Item	Check	
Signage		
The position of the designated accessible car parking bays is clearly signed from the entrance to the car park	✓	✗
Appropriate signage (symbol) is provided at ground level (white or yellow and at least 1400 mm high)	✓	✗
Appropriate signage (text and symbol) is provided vertically at the back of each bay which is high enough to be seen when the accessible bay is occupied	✓	✗
Position of bays		
The accessible parking bays are sited as close to the main entrance of the building as is practicable	✓	✗
The travel distance from the accessible bays to the main entrance (or the accessible entrance) is less than 50 metres	✓	✗
If the travel distance is greater than 50 metres, the route is undercover	✓	✗
Circle ✗ if the travel distance is greater than 100 metres		✗
The bays are positioned appropriately relative to the slope, if any, of the site between the accessible bay and the main accessible entrance	✓	✗
Number of bays		
The number of bays provided is appropriate to the use of the building or buildings	✓	✗
Size of bays		
If a single accessible parking bay is provided, it is a minimum of 4800 mm long by 3600 mm (including a 1200 mm transfer zone)	✓	✗
or		
If a bank of accessible parking bays is provided each bay is at least 4800 mm long by 2400 mm wide with an additional transfer zone of 1200 mm between bays	✓	✗
The minimum size of standard parking bays is at least 4800 mm by 2400 mm	✓	✗
There is a suitable marked zone at the end of each bay to allow entry and egress from the rear of a vehicle	✓	✗
Pedestrian routes		
Routes through the car park are clearly signed	✓	✗
Routes through the car park are clearly defined using texture or colour	✓	✗
Routes through the car park are wide enough	✓	✗
Routes through the car park are free from hazards	✓	✗
Routes through the car park are slip resistant (including when wet)	✓	✗
Routes through the car park are likely to be free from puddles when wet	✓	✗
Profile paving (blister) is provided at dropped kerbs	✓	✗

Item	Check	
Surface finish		
The surface finish to all car parking areas and setting down points (if any) is firm	✓	✗
The surface finish to all car parking areas and setting down points (if any) is well maintained	✓	✗
Open joints between paving slabs (if any) do not exceed 10 mm	✓	✗
Maximum difference in level of paving slabs (if any) does not exceed 5 mm	✓	✗
Lighting		
Lighting is provided to the car park	✓	✗
Lighting is provided to the setting down point (if any)	✓	✗
Lighting is well maintained	✓	✗
Lighting, in terms of illuminance and provision, is appropriate	✓	✗
Setting down point		
A setting down point at least 6600 mm long is provided as close as practicable to the main accessible entrance	✓	✗
The setting down point is appropriately signed	✓	✗
There is a level kerb	✓	✗
A shelter is provided	✓	✗
There is an appropriate procedure in place to manage the use of the setting down point	✓	✗
Management		
There are appropriate procedures in place to manage the accessible parking bays and setting down points	✓	✗
Ticket machines (if any) are fully accessible to disabled people	✓	✗
Generally		
Methods of controlling entry to the car park are suitable for all potential disabled users	✓	✗

External areas checklist

Item	Check	
Footpaths		
Pathways are clearly defined	✓	✗
Surfaces to pathways are firm	✓	✗
Circle ✗ if loose gravel is provided as the surface finish		✗
Circle ✗ if cobble stones or similar are provided as the surface finish		✗
Surfaces are slip resistant in all weather conditions	✓	✗
The colour of the finish is uniform throughout	✓	✗
Circle ✗ if there are changes in colour which could appear as steps to people with poor vision		✗
Open gaps in the surface finish (if any) are less than 10 mm wide	✓	✗
Maximum difference in level of surfaces (if any) does not exceed 5 mm	✓	✗
If the level of the path is higher than the adjacent ground: there is a kerb of minimum height 100 mm provided which does not present a tripping hazard to users	✓	✗
or		
there is a tapping rail of minimum height 100 mm provided which does not present a tripping hazard to users	✓	✗
The kerb or tapping rail is contrasted in terms of colour and luminance with the path and the adjacent ground	✓	✗
Width of footpaths		
The path is 1200 mm wide minimum	✓	✗
or		
The path is 1500 mm wide minimum	✓	✗
The path is 1800 mm wide minimum	✓	✗
Gradient of footpath		
The gradient of the path is 1:20 or flatter	✓	✗
The cross fall gradient of the path is 1:50 or flatter	✓	✗
Changes of direction		
Corners at changes of direction of the path are splayed or rounded	✓	✗
Gratings		
Slots in gratings (if any) do not exceed 13 mm wide	✓	✗
Circular holes in gratings (if any) do not exceed 18 mm diameter	✓	✗
Drainage channels at dropped kerbs (if any) are protected with a flat plate across the channel for the full length of the dropped kerb	✓	✗
Dropped kerbs		
The width of the dropped section of kerb is 1200 mm minimum	✓	✗

Item	Check	
or		
The width of the dropped section of kerb is 2000 mm minimum	✓	✗
The gradient at dropped kerbs is 1:15 or flatter	✓	✗
The appropriate use of tactile paving is made (in terms of colour, profile, size and design)	✓	✗
Handrails (see also Stairs checklist)		
A suitable handrail (or balustrade) is provided at all changes of level	✓	✗
Lighting		
The footpaths are provided with lighting	✓	✗
The lighting is provided which does not create areas of uneven lighting or strong shadows	✓	✗
The lighting provided enhances the colour and luminance definition of any potential hazards	✓	✗
Street furniture		
Bollards (if provided) are a minimum 1000 mm high	✓	✗
Bollards (if provided) are adequately contrasted in terms of colour and luminance with the background against which they will be viewed:		
for the full height	✓	✗
or		
for a minimum of 150 mm at the top of the bollard	✓	✗
Bollards used for car parking or to control entrances (if any) recess to be flush with the surface of the footpath or area when unlocked	✓	✗
All potential hazards (litter bins, planters, seats, signs, etc.) are logically placed	✓	✗
All potential hazards are contrasted in terms of colour and luminance with the background against which they will be viewed:		
for the full height	✓	✗
or		
for a minimum or 150 mm at the top of the potential hazard	✓	✗
The minimum distance between pieces of street furniture is 1000 mm	✓	✗
For long or sloping journey routes, appropriate seating is provided every 100 metres	✓	✗
Projections		
Plants, bushes and shrubs are managed such that they do not extend or project into the circulation route	✓	✗
There is a minimum 2100 mm clear headroom to any trees, bushes or other objects such as awnings, bay windows, etc. (if any) projecting into the line of the footpath	✓	✗
Projections below 2100 mm are protected with an upstand or tapping rail at least 100 mm high	✓	✗
There is clear evidence of appropriate management to the paths and the adjacent areas	✓	✗

Entrances and reception areas checklist

Item	Check	
The entrance		
The principal entrance to the building is accessible to everyone	✓	✗
The entrance is clearly distinguishable when approaching the building	✓	✗
A canopy providing adequate shelter is provided over the entrance doors	✓	✗
or		
The entrance doors are recessed	✓	✗
Transitional lighting is provided for people entering the building	✓	✗
Entry systems (other than those controlling car parking)		
Entry systems are suitable for use by disabled people	✓	✗
The entry system incorporates an induction coupler	✓	✗
The entry system incorporates an LED or similar text display	✓	✗
Swipe card entry systems are appropriately positioned 750 mm to 1000 mm affl	✓	✗
The position of the entry system is logical	✓	✗
The position of the entry system is clearly identified using colour and luminance contrast	✓	✗
The top operational button of the entry is 1200 mm affl or less	✓	✗
Appropriately designed signage is provided with instructions of how to use the entry system	✓	✗
Appropriately designed signage is provided with instructions of how to get assistance if a disabled person cannot use the entry system	✓	✗
Entrance doors		
The minimum clear opening width through one leaf is 800 mm	✓	✗
There is a 300 mm clear space beside the leading edge	✓	✗
The maximum pressure needed to open the doors is 20 N	✓	✗
Delayed action closers are provided	✓	✗
If the doors open outwards they are:		
recessed	✓	✗
or		
the swing area is protected adequately to prevent collisions	✓	✗
Vision panels with a visibility zone extending between 500 mm and 1500 mm affl are provided	✓	✗
Where necessary, manifestation is provided at 1050 mm and 1500 mm affl	✓	✗
If provided, manifestation is effective at all times the building is in use	✓	✗

Item	Check	
If glass doors are provided:		
the frame contrasts with the surrounding wall/screen	✓	✗
or		
the presence of the doors is differentiated from the rest of the wall/screen	✓	✗
Edges of doors are clearly visible when held in the open position	✓	✗
Door furniture is distinguishable in terms of colour and luminance contrast with the door	✓	✗
Door handles can be reached, gripped and used with minimum effort	✓	✗
A kicking plate at least 400 mm deep is provided	✓	✗
Automatic doors		
Automatic doors, either manually or automatically operated, are provided	✓	✗
Warning of the presence of the automatic doors is provided	✓	✗
Warning of the direction of opening of the automatic doors is provided	✓	✗
Opening and closing door speed is appropriate	✓	✗
Doors remain open for an appropriate time	✓	✗
Appropriate safety arrangements are provided to prevent doors closing if there is an obstruction	✓	✗
Revolving doors		
Circle ✗ if revolving doors are provided		✗
If a revolving door is provided, there is an alternative entrance with a clear opening width of 800 mm adjacent and clearly visible	✓	✗
Circle ✗ if the alternative door is not provided with an automatic opening device		✗
Circle ✗ if the alternative door is not operational at all times when the revolving door is in use		✗
Thresholds		
The threshold is:		
flush	✓	✗
or		
the maximum change of level for the threshold is 15 mm	✓	✗
Doormats		
Circle ✗ if coir or other deep pile matting is provided		✗
Mats are ribbed and capable of supporting the weight of a wheelchair	✓	✗
Mats are recessed	✓	✗
Circle ✗ if mats present a tripping hazard		✗
Circle ✗ if mats are loose or not firmly fixed		✗
Mats are of sufficient size to allow at least one revolution of the wheel of a wheelchair to pass over	✓	✗

Entrances and reception areas checklist (continued)

Item	Check	
Lobbies		
Lobbies, if any, are appropriately designed to allow easy access and egress for all users	✓	✗
Exits		
All exits have the same level of accessibility as that described for Entrances above	✓	✗
Reception areas: approach		
The approach to the reception area is smooth and level	✓	✗
Directional signage to identify the position of the reception area is appropriate in terms of provision and design	✓	✗
Reception areas: seating		
A variety of seating in terms of seat height and with and without arms is provided	✓	✗
Seating is sufficiently robust to allow someone to use the arms as assistance when sitting or standing	✓	✗
Spaces are provided or can be created which allow wheelchair users to sit within the main seating area	✓	✗
Seating contrasts in terms of colour and luminance with all backgrounds against which it will be viewed	✓	✗
Reception areas: lighting		
Lighting to the reception area is appropriate to allow easy communication and comfort for users	✓	✗
Reception desks		
The reception desk is logically placed	✓	✗
The position of the reception desk is identifiable from the entrance to the building	✓	✗
There is sufficient manoeuvring space in front of the reception desk	✓	✗
A lowered section is provided to the reception desk	✓	✗
An adequate knee recess is provided to enable form filling at the seated position	✓	✗
Communication is possible across the desk from the standing or seated position	✓	✗
An induction loop is provided	✓	✗
If provided, the induction loop is operational at the time of the audit	✓	✗
If necessary, the completion of forms for all users is possible	✓	✗
Lighting at the desk is appropriate to allow the face of the receptionist to be clearly seen and	✓	✗
sufficient to enable the completion of any forms, etc.	✓	✗
Illuminance at the desk is controllable	✓	✗
Circle ✗ if a reflective glass screen is provided		✗
There is evidence of staff training in disability awareness and clear lip speaking	✓	✗
Instructions are available in alternative formats	✓	✗

Horizontal circulation checklist

Item	Check	
Corridors and passageways		
Corridors have a minimum clear width of 1200 mm	✓	✗
If a building is frequently used by a significant number of wheelchair users the corridors are:		
1800 mm wide	✓	✗
or		
have appropriately placed passing places	✓	✗
In existing buildings where corridor widths are less than 1200 mm:		
widening internal door openings is being considered	✓	✗
and		
consideration is being given to the provision of passing places	✓	✗
Circle ✗ if doors (other than to accessible toilets) open into the corridors		✗
Corridors are splayed or rounded at corners	✓	✗
Surfaces		
The floor surface finishes used are firm enough to be suitable for wheelchair users	✓	✗
Carpets (if provided) have a shallow, dense pile	✓	✗
The floor surface is matt or low reflectivity	✓	✗
Floor surfaces are slip resistant, especially when wet	✓	✗
Circle ✗ if the floor surface appears shiny and potentially 'slippery'		✗
The pattern of any floor covering is plain or with a subtle pattern	✓	✗
Junctions between finishes are level and firmly fixed	✓	✗
Appropriate contrast, in terms of colour and luminance, is provided at the junction of the floor with the wall	✓	✗
Wall surfaces are smooth to the touch	✓	✗
Circle ✗ if the surfaces to the floor or walls adversely affect the acoustics		✗
Glazed walls, screens and doors		
Appropriately designed manifestation is provided to glass screens at 1050 mm and 1500 mm affl	✓	✗
Handrails (see also Stairs and steps, internal and external checklist)		
A handrail is provided along the length of the corridors	✓	✗
Internal doors		
The minimum clear opening width of all internal doors is 750 mm	✓	✗

Horizontal circulation checklist (continued)

Item	Check	
If double doors are provided, the minimum clear opening width of one leaf of the doors is 750 mm	✓	✗
There is a minimum 300 mm clear space beside the leading edge	✓	✗
For doors on circulation routes: the swing area is protected	✓	✗
or		
the doors are recessed	✓	✗
All door furniture (handles, kicking plates and finger plates) are distinguishable in terms of colour and luminance contrast from the door	✓	✗
Door handles are lever style with a return to the door on the open end	✓	✗
A 400 mm deep kicking plate is provided	✓	✗
The kicking plate is fixed with non-projecting fixings	✓	✗
The pressure required to open an internal door is less than 20 N	✓	✗
Delayed action door closers are provided	✓	✗
The delayed action door closers, if any, are operating correctly	✓	✗
Internal doors are clearly identifiable in terms of colour and luminance contrast with the surrounding wall	✓	✗
The leading edge of all internal doors is contrasted in terms of colour and luminance and is clearly visible when open	✓	✗
A vision panel is provided in internal doors which has a zone of visibility between 500 mm and 1500 mm affl	✓	✗

Fire doors
Fire doors:

	Check	
can be opened easily by disabled people	✓	✗
or		
are held open on electrically powered hold open devices linked to the fire alarm system	✓	✗
Checks have been undertaken to ensure fire doors held open on electrically powered open devices (or similar) can be opened by disabled people when activated to the closed position	✓	✗

Lobbies

	Check	
Internal lobbies, if provided, are appropriately designed to allow easy manoeuvrability	✓	✗

Lighting

	Check	
Illuminance in corridors is a minimum of 100 lux	✓	✗

Stairs and steps, internal and external checklist

Item	Check	
General information		
The stairs are made up of a straight flight or flights of stairs	✓	✗
Circle ✗ if the stairs are spiral or contain a section which has winders		✗
Circle ✗ if the stairs have open risers		✗
Circle ✗ if the stairs (including any landings) are not protected to the underside to prevent a user colliding with the stair		✗
There are no more than 12 risers in each flight	✓	✗
The unobstructed width of the stairs is a minimum 1000 mm (between handrails if any)	✓	✗
Corduroy pattern tactile flooring, which extends the width of the flight, is provided at the top and bottom of each flight	✓	✗
There is a change in floor texture at the top and bottom of each flight that gives adequate warning of the presence of the stairs	✓	✗
Design, size and layout		
There is a clear landing of at least 1200 mm at the top of each flight	✓	✗
There is a clear landing of at least 1200 mm at the bottom of each flight	✓	✗
The risers on each step are the same dimension	✓	✗
The treads on each step are the same dimension	✓	✗
The height of the risers is between 150 mm and 170 mm	✓	✗
The going of the tread is between 250 mm and 300 mm	✓	✗
Circle ✗ if there are projected nosings on the steps		✗
Handrails		
A handrail is provided to each side of the stair	✓	✗
On a wide flight of stairs, a central handrail is provided	✓	✗
The handrail is between 40 mm and 50 mm diameter or	✓	✗
The handrail is oval with dimensions between 50 mm wide and 38 mm deep	✓	✗
The handrail can be gripped along its full length	✓	✗
There is a clear space of at least 50 mm between the handrail and the adjacent wall	✓	✗
The fixing to the handrail allows the hand gripping the rail to pass by the fixing without the fixing making contact with or injuring the user's hand	✓	✗
The handrail extends horizontally at least 300 mm beyond the first and last nosing in the flight	✓	✗

Stairs and steps, internal and external checklist (continued)

Item	Check	
The handrail is continuous around landings	✓	✗
The top of the handrail is 900 mm to 1000 mm (measured vertically) above the nosings	✓	✗
The top of the handrail is 900 mm to 1100 mm (measured vertically) above landings	✓	✗
The material that the handrail is made of/covered with is timber or nylon or other material that is easy and comfortable to grip, smooth and not cold to the touch	✓	✗
Tactile information, indicating floor levels, etc., is given on the handrails	✓	✗
Contrast The nosings on each step are adequately contrasted in terms of colour and luminance	✓	✗
The contrast extends 40 mm on the tread and the riser	✓	✗
The contrast to the nosings can be seen when ascending and descending the stairs	✓	✗
The handrails are adequately contrasted in terms of colour and luminance with the background against which they will be viewed	✓	✗
There is a tactile warning surface or a change of floor colour and texture at the head and foot of the stair that gives adequate warning of the presence of the stairs	✓	✗
Lighting Illuminance at tread/floor level is a minimum of 200 lux	✓	✗
Circle ✗ if the lighting provided is in the risers of the steps		✗
Circle ✗ if the lighting to the stairs is not even, causes glare, or is disorientating to users		✗
Maintenance The handrails are securely fixed to the supporting wall	✓	✗
Nosings are in good order	✓	✗
Circle ✗ if nosings present a tripping hazard or are in poor repair		✗
The light fittings to the stairs are clean	✓	✗
Additional for external steps only Circle ✗ if an 800 mm deep corduroy tactile warning surface is not provided at the top and bottom of each flight		✗

Ramps, internal and external checklist

Item	Check	
Slope		
The travel distance on the slope is less than 10 metres	✓	✗
The slope rises vertically less than 500 mm over its length	✓	✗
If the slope is part of a series of ramps the total vertical rise is less than 2 metres	✓	✗
If the gradient does not exceed 1:20, the length of the ramp is less than 10 metres	✓	✗
If the gradient does not exceed 1:15, the length of the ramp is less than 5 metres	✓	✗
If the gradient is 1:12, the length of the ramp is less than 2 metres	✓	✗
Circle ✗ if any part of the ramp is steeper than 1:12		✗
Width		
The minimum surface width of the ramp is 1200 mm	✓	✗
If the ramp is 1200 mm wide, it is shorter than 5 metres	✓	✗
The minimum unobstructed clear width is 1000 mm	✓	✗
If the surface width of the ramp is less than 1800 mm, there is a clear unobstructed view along the whole length of the ramp	✓	✗
Landings		
For a ramp that does not exceed 1:20, there is a level landing every 10 metres	✓	✗
For a ramp that does not exceed 1:15, there is a level landing every 5 metres	✓	✗
If a landing is provided, it is at least 1500 mm long	✓	✗
If a passing place is needed, a landing of at least 1800 mm by 1800 mm is provided	✓	✗
There is a level landing at least 1200 mm long at the top of the ramp	✓	✗
There is a level landing at least 1200 mm long provided at the bottom of the ramp	✓	✗
Circle ✗ if the landings at the top and bottom of the ramp could be encroached upon by the swing of a door	✓	✗
There is a level landing provided at each change of direction of the ramp	✓	✗
If uncovered, the cross fall to the landings is less than 1:50	✓	✗
All landings will not allow water to stand on the surface	✓	✗
Surfaces		
The surface to the ramp is:		
smooth	✓	✗
and		
firm enough to take the load imposed on it	✓	✗
and		
slip resistant when wet	✓	✗
and		
easy to maintain	✓	✗

Ramps, internal and external checklist (continued)

Item	Check	
If different surface finishes are used on the ramp, landings and approach paths, coefficients of friction of all surfaces are similar	✓	✗
Circle ✗ if a tactile warning has been provided at the top or bottom of the ramp		✗
Circle ✗ if there is a surface pattern to the ramp (cross stripes, etc.) that could visually appear as steps		✗
Handrails		
A handrail is provided on the ramp	✓	✗
A handrail is provided on both sides of the ramp	✓	✗
If a very wide ramp is provided, there is a central handrail	✓	✗
The handrail is between 40 mm and 50 mm diameter or The handrail is oval with dimensions between 50 mm wide and 38 mm deep	✓	✗
The handrail can be gripped along its full length	✓	✗
There is a clear space of at least 50 mm between the handrail and the adjacent wall (if any)	✓	✗
The fixing to the handrail allows the hand gripping the rail to pass by the fixing without the fixing making contact with or injuring the user's hand	✓	✗
The handrail extends horizontally at least 300 mm beyond the start and finish of the ramp	✓	✗
The handrail is continuous around landings	✓	✗
The top of the handrail is 900 mm to 1000 mm (measured vertically) above the surface of the ramp	✓	✗
The top of the handrail is 900 mm to 1100 mm (measured vertically) above surface of the landings (if any)	✓	✗
The material that the handrail is made of/covered with is timber or nylon or other material that is easy and comfortable to grip, smooth and not cold to the touch	✓	✗
Alternative steps		
Circle ✗ if no steps are provided as an alternative to using the ramp		✗
Lighting		
Illuminance on the floor surface of the ramp is at least 200 lux	✓	✗
Light fittings are well maintained	✓	✗
Circle ✗ if the lighting to the ramp is not even, causes glare or is disorientating to users		✗

Item	Check	
Temporary ramps		
Circle ✗ if it is possible to provide a permanent ramp		✗
The surface of the temporary ramp is at least 800 mm wide	✓	✗
The surface is slip resistant and well drained	✓	✗
There is an upstand which prevents wheels slipping over the edge	✓	✗
The maximum gradient is 1:12	✓	✗
The ramp is appropriately contrasted in terms of colour and luminance with the background against which it will be viewed	✓	✗
The ramp is well illuminated when in place	✓	✗

For external steps see Stairs, internal and external checklist.

Lifts checklist

Item	Check	
Passenger lifts		
Manoeuvring		
There is a manoeuvring space of at least 1500 mm by 1500 mm outside the entrance to the lift	✓	✗
The entrance door to the lift has a clear opening width of at least 800 mm	✓	✗
or		
In a building which may be used by sports wheelchairs, the entrance door has a clear opening width of at least 1000 mm	✓	✗
The minimum internal dimensions of the lift at floor level are 1100 mm by 1400 mm	✓	✗
Lift cars meet level with floors at all floors served by the lift	✓	✗
The floor surface inside the lift is firm and slip resistant	✓	✗
A handrail is provided on three sides inside the lift	✓	✗
If a handrail is provided, it can be gripped (40 mm to 50 mm diameter)	✓	✗
If a handrail is provided, it is located at 900 mm affl	✓	✗
Calling the lift		
The position of the buttons to call the lift is logical	✓	✗
The call panel is clearly distinguishable in terms of colour and luminance contrast from its background	✓	✗
The buttons illuminate when pressed	✓	✗
The information on the buttons is appropriately embossed	✓	✗
The buttons are placed between 900 mm and 1200 mm affl	✓	✗

Lifts checklist (continued)

Item	Check	
Lift arrival indication is given (using a text or a symbol)	✓	✗
An audible sound indicates the arrival of the lift	✓	✗
Circle ✗ if, in a bank of lifts, it is not clear which lift has arrived		✗

Using the lift

The lift doors remain open for at least 5 seconds	✓	✗
The door reactivating device operates on infrared or photo eye sensors	✓	✗
Circle ✗ if the reactivating device operates on a pressure sensor		✗
Audible warnings of the doors opening and closing are provided	✓	✗
The volume and clarity of the message is appropriate	✓	✗
Floor level indicators can be seen when the lift is full	✓	✗
Outside the lift, tactile and visual floor level indicators are provided which are clearly visible to users when the lift door opens	✓	✗
There is a mirror on the rear wall of the lift (above handrail level)	✓	✗
The mirror is of sufficient size to allow a wheelchair user to get information when reversing out of the lift	✓	✗
The mirror is of sufficient size to allow a wheelchair user to see a floor indicator placed above the door	✓	✗
The floor level arrived at is announced audibly and visually	✓	✗

Controls within the lift

The control panel is situated logically within the lift	✓	✗
The buttons are within easy reach by all users	✓	✗
Call or control buttons are located between 900 mm and 1100 mm affl	✓	✗
Call or control buttons are located at least 400 mm away from any return wall	✓	✗
Call buttons are 20 mm to 30 mm diameter (or equiv. if square or rectangular)	✓	✗
Information on the call buttons is provided in Braille	✓	✗
Information on the call buttons is embossed (1 mm)	✓	✗
Call buttons illuminate when pressed	✓	✗
Circle ✗ if the call buttons are touch sensitive		✗

Contrast and lighting

The position of the lift is adequately identified using colour and luminance contrast and is well illuminated	✓	✗
Lighting within the car is a minimum 100 lux	✓	✗

Item	Check	
Emergency equipment		
There is an alarm button fitted which is within easy reach for wheelchair users	✓	✗
Information of what to do in an emergency is provided	✓	✗
The information is provided in large print (sans serif font, min 12 point)	✓	✗
The information is also provided in Braille	✓	✗
There is an audible indication that assistance has been requested	✓	✗
There is a visual indication that assistance has been requested	✓	✗
If the emergency system relies on audible communication, it is supported by written text information explaining emergency procedures	✓	✗
or		
a text phone	✓	✗
If a telephone is provided, an inductive coupler is fitted	✓	✗
Generally		
The surface finishes used within the lift are not highly reflective	✓	✗
The surface finishes used within the lift do not create an unacceptable acoustic environment	✓	✗
Platform lifts		
If a platform lift is provided, compliance with the requirements of BS 6440:1999 has been checked	✓	✗
The platform lift has no enclosure, there is no floor penetration and the rise is no greater than 2 metres	✓	✗
The platform lift has an enclosure and the rise is no greater than 4 metres	✓	✗
The minimum clear dimensions of the platform are 1050 mm wide by 1250 mm long	✓	✗
Wheelchair stair lifts		
If a wheelchair stair lift is provided, compliance with the requirements of BS 5776: 1996 Annex A has been checked	✓	✗
When parked, the stair lift leaves a clear unobstructed width on the stairs	✓	✗
When installed, the minimum clear width between the folded down platform and the handrail opposite is 600 mm	✓	✗
Circle ✗ if the installation of the platform lift results in the loss of one handrail on the stairs		✗
A means of summoning assistance is provided	✓	✗
Any alarms provided comply with the requirements of ISO 9386-2	✓	✗

WC facilities (standard and accessible) checklist

Item	Check	
Ambulant accessible WC		
Number of compartments		
There is at least one ambulant accessible compartment in each standard male WC accommodation	✓	✗
There is at least one ambulant accessible compartment in each standard female WC accommodation	✓	✗
General spatial layout		
The internal dimensions of the compartment are at least 800 mm wide and 1500 mm long	✓	✗
Doors		
The door to the cubicle opens outwards	✓	✗
or		
If the door opens inwards, there is a minimum 450 mm diameter of space between the door swing and any fitting within the WC	✓	✗
If the door opens inwards, it can be opened outwards or removed easily in an emergency	✓	✗
WC		
The WC is placed centrally across the width of the WC	✓	✗
The height of the seat is at 480 mm affl	✓	✗
The flush is a 'spatula' type	✓	✗
Grab rails		
There are two horizontal grab rails, each 600 mm long, placed 680 mm affl and with their centre line 650 mm from the rear wall of the cubicle	✓	✗
There is a 600 mm grab rail placed vertically on one side wall. The bottom of the rail is 700 mm affl and the rail is positioned 200 mm in front of the WC pan	✓	✗
Grab rails protrude less than 90 mm into the cubicle space	✓	✗
If a urinal is provided for use by ambulant disabled people, a 500 mm grab rail is provided to both sides of the urinal	✓	✗
Lighting		
Lighting provides a minimum 100 lux	✓	✗
Circle ✗ if the lighting produces glare		✗
Other essential items		
A coat hook is provided at between 1200 mm and 1400 mm affl	✓	✗
The is evidence of appropriate, ongoing management for the day-to-day use of the facility (e.g. refilling, replacing essential items regularly)	✓	✗

Item	Check	
Wheelchair accessible WC		
Accessibility		
Signage indicating the route to, and position of, the toilet facilities is adequate	✓	✗
Circle ✗ if a unisex accessible toilet that can be accessed independently of other toilet accommodation is NOT provided		✗
Circle ✗ if entry into the accessible toilet is controlled by a RADAR key or some other method of locking		✗
General spatial layout		
Accessible WC clear floor area is 2200 mm by 1500 mm	✓	✗
The travel distance to an accessible WC with an appropriate transfer side is less than 100 metres if on the same level	✓	✗
or		
40 metres if the travel distance includes using a lift	✓	✗
In employment situations, an assessment has been made of the needs of the disabled employee/s to determine the suitability of the above dimensions	✓	✗
Doors		
Door opens outwards	✓	✗
or		
Door opens inwards	✓	✗
Door can be removed easily in an emergency	✓	✗
WC		
The centre line of the WC is 500 mm from the nearest wall	✓	✗
The space between the WC and the wall is kept clear to allow a carer to assist if required	✓	✗
The height to the top of the seat is 480 mm affl	✓	✗
The seat is suitable for heavy duty use (not a gap-front style)	✓	✗
The front of the WC projects 750 mm from the face of the supporting wall	✓	✗
The seat is securely fixed and of a good quality	✓	✗
There is a low-level cistern that could be used as a backrest	✓	✗
or		
A backrest with padded section has been provided 680 mm affl	✓	✗
The flush is a 'spatula' type	✓	✗
The flush is on the open transfer side	✓	✗
There is a single-sheet toilet paper dispenser within easy reach of the WC	✓	✗
There is a dispenser for hand wipes within easy reach of the WC	✓	✗

WC facilities (standard and accessible) checklist (continued)

Item	Check	
Grab rails		
There is a drop-down rail that is easy to operate from the seated position placed 320 mm from the centre line of the WC	✓	✗
There is a vertical grab rail that extends between 800 mm and 1400 mm affl and is placed to the open side of the WC, its centre line 470 mm from the centre of the WC	✓	✗
There is a grab rail on the wall adjacent to the WC that is 600 mm long and is placed 680 mm affl. One end of the grab rail is 200 mm away from the wall supporting the WC	✓	✗
There is a grab rail on the door at a height that will enable someone using a wheelchair to pull the door closed	✓	✗
There are two vertical grab rails, one either side of the wash hand basin, which extend between 800 mm and 1400 mm affl	✓	✗
The diameter of all grab rails is between 32 mm and 35 mm and they are easy to grip even when wet	✓	✗
Washing and drying hands		
There is a wall supported wash hand basin (not pedestal supported) provided with the top of the basin 720 mm to 740 mm affl	✓	✗
The basin can be reached and used easily whilst seated on the WC	✓	✗
The basin is 140 mm to 160 mm away from the WC	✓	✗
A single-lever mixer tap is provided	✓	✗
A soap dispenser is provided	✓	✗
The temperature of the water being delivered from the tap is controlled	✓	✗
The temperature of the water is appropriate	✓	✗
There is a paper towel dispenser that can be reached whilst seated on the WC	✓	✗
A manually operated (not proximity operated) warm-air dryer with the activation button max 1200 mm affl is provided adjacent to the hand basin	✓	✗
Lighting		
A white pull cord for the light is provided adjacent to the door or	✓	✗
A device is installed which automatically switches on the light when the toilet door is opened	✓	✗
Lighting provides a minimum 100 lux	✓	✗
Circle ✗ if the lighting produces glare		✗

Item	Check	
Alarm		
An emergency alarm call system is provided	✓	✗
A red alarm pull is fitted	✓	✗
The alarm pull is provided with two red bangles which are either:		
50 mm diameter	✓	✗
or		
open triangular with 50 mm sides	✓	✗
One bangle is positioned 100 mm affl and one between 800 mm and 1000 mm affl	✓	✗
The alarm is both audible and visual within the WC when activated	✓	✗
There is a reset button provided which is clearly visible, signed and positioned such that it can be reached whilst seated in a wheelchair	✓	✗
Colour contrast		
The sanitary fittings and facilities provided contrast with their background adequately in terms of colour and luminance	✓	✗
The contrast at the junction of the floor with the walls, in terms of colour and luminance, is adequate	✓	✗
Other essential features		
The floor covering is slip resistant (including when wet)	✓	✗
A small shelf for use by colostomy users is provided	✓	✗
Coat hooks are provided at 1050 mm and 1400 mm affl	✓	✗
A mirror is provided above the hand basin	✓	✗
A shaver point is provided adjacent to the mirror and between 800 mm and 1000 mm affl	✓	✗
If provided in the standard toilet accommodation, vending machines are provided in the accessible toilet	✓	✗
or		
They are accessible in the standard toilets or reasonably in another place	✓	✗
Wall tiles or finishes are non-reflective	✓	✗
Facilities for disposable items are provided in a manner that does not impinge upon the manoeuvring space within the WC	✓	✗
There is evidence of appropriate, ongoing management for the day-to-day use of the facility (e.g. refilling, replacing essential items regularly)	✓	✗

WC facilities (standard and accessible) checklist (continued)

Item	Check	
Standard WC facilities		
Generally (in all cases)		
Signage indicating the route to, and position of, the WC facilities is adequate	✓	✗
Contrast, in terms of colour and luminance, of all fittings with the background against which they will be viewed is appropriate	✓	✗
The contrast at the junction of the floor with the walls, in terms of colour and luminance, is adequate	✓	✗
Wall tiles or finishes are non-reflective	✓	✗
Lever mixer taps are provided	✓	✗
The temperature of the water being delivered from the tap is controlled	✓	✗
The temperature of the water is appropriate	✓	✗
The floor covering is slip resistant (including when wet)	✓	✗
Fitting such as flushes to WC and light switches are usable by people with restricted dexterity	✓	✗
Urinals are provided at a choice of heights affl	✓	✗
Facilities for washing and drying hands are reachable by all potential users	✓	✗
Lighting provides a minimum of 100 lux	✓	✗
The is evidence of appropriate, ongoing management for the day-to-day use of the facility (e.g. refilling, replacing essential items regularly)	✓	✗

Appendix B
Sample audits

The following examples are extracts from sample audit reports and are intended to illustrate various report formats. The extracts from the Michael Sarna House and Lydia House reports use a combination of narrative and tabular formats. The extract from the Oliver House report illustrates a tabular format. The Michael Sarna House example includes an extract from a typical Design Guidance Manual. A manual such as this can be attached to an audit report to give design guidance and is particularly useful where there are a number of buildings being assessed. The manual can include photographs or drawings to illustrate the guidance.

Michael Sarna House, common areas, office building

Sanitary facilities

Accessible toilet (entrance level)

The dimensions of the accessible toilet are 1430 × 2140 mm, which is less in both directions than that recommended in BS 8300 (1500 mm × 2200 mm). The width of the toilet as provided (1430 mm) is also less than the minimum dimensions stated in the current Part M of the Building Regulations (1500 mm). Whilst this may not appear to be a significant difference, it is less than the requirements in force when this 2-year-old building was constructed.

The effect is that there is a reduction in the usable space and insufficient wheelchair turning space for some users.

In addition, because of space constraints outside the toilet, the door opens inwards. This reduces still further the usable floor area in the toilet and, if someone falls against the door, would impede helpers entering the toilet to give assistance. The sign is positioned too high and displays the international disabled sign of a wheelchair user but does not indicate if this facility is actually a toilet. The kick plate is less than the recommended 400 mm and evidence already exists of damage to the door because of this.

The lever handle and turn lock are suitable. The existing grab bars are appropriate and the sink is accessible from the toilet as is the soap dispenser (Figure B.1).

The existing toilet paper dispenser is not of a recommended type. No hand towel facility is provided. There is insufficient colour contrast generally and there is no alarm to call assistance in an emergency.

There are no other accessible toilet facilities in the building, making travel distances from most of the tenanted areas to an accessible facility unacceptably long. At the time of the audit, it was noted that some of the standard toilets could be adapted to include an accessible facility.

Figure B.1

We would strongly recommend that a survey is undertaken of the sanitary accommodation in the building and appropriate areas are identified for upgrading in the future. In our opinion, there is currently no toilet facility within this building that can be considered suitable for many disabled people to use and certainly not one that can be considered as meeting acceptable good practice in terms of accessibility.

Standard toilets are provided with twist-style taps. When necessary, perhaps in response to a workplace assessment for an employee or when the taps are replaced, twist taps should be replaced with lever taps.

Extract of audit recommendations

An extract of the audit recommendations for sanitary facilities is shown in Table B.1.

Table B.1 Michael Sarna House access audit recommendations

Para no.	DGM Para no.	Problems identified	Remedial work required	Priority	Linked to maintenance schedule	Cost (£)
Standard toilet facilities						
8.0	11.0	No directional signage to toilets throughout building	Provide new signage in accordance with DGM recommendations	A	Yes	250
		No projecting signage outside toilets		A	Yes	100
		Toilet door, pictogram signage only		A	Yes	50
8.0	10.12	Lever taps – general recommendations	As taps fail replace with lever taps	B	Yes	As reqd
8.0	12.7	Ceiling light and extractor fan in first floor shower room	Ensure that all fittings are cleaned and maintained on a regular basis	B	Yes	0
Accessible toilets						
8.1	11.1, 11.2, 11.4 & 11.9	Lack of directional and projecting signage for accessible toilets	Provide new signage in accordance with DGM recommendations	A	Yes	0
		Pictogram door signage only		A	Yes	0
Ground floor accessible toilet						
8.1	10.9	Position of hand drier	Reposition so hand drier can be used when sitting on WC	A	Yes	50
8.1	8.3 & 10.10	Poorly contrasted sanitary fittings	Improve contrast where appropriate	B	Yes	Choice
8.1	10.13	No emergency pull cord	Provide emergency pull cord in accordance with DGM recommendations	A	Yes	300

Para no.	DGM Para no.	Problems identified	Remedial work required	Priority	Linked to maintenance schedule	Cost (£)
8.1	10.7	Turn lock difficult to operate	Change lock	A	Yes	50
8.1	10.6	Cluttered cubicle area	Ensure that cubicle area is kept free from any obstructions	B	Normal working procedures	0
8.1	N/A	No false ceiling in cubicle	Provide ceiling similar to standard toilet areas	B	Yes	350

Second floor accessible toilet. N.B. This is currently unsuitable as an accessible toilet

8.1	N/A	Second floor accessible toilet is not suitable	Remove existing signage for accessible toilet on second floor and replace with standard toilet sign	A	Yes	0

If toilet is to be made fully accessible

8.1	N/A	Sink in cubicle	Remove to create more space	A	Yes	100
8.1	10.13	No emergency pull cord	Provide emergency pull cord in accordance with DGM recommendations	A	Yes	300
8.1	10.9	Position of drying facilities	Reposition so a person can dry their hands whilst sitting on WC	A	Yes	50
8.1	10.7	Door bolt difficult to operate	Change to lever type lock or equivalent	A	Yes	50

Extracts from Design Guidance Manual

Contents (example)   Page (example)

1.1   Designated accessible parking

The number of accessible parking places provided should be appropriate for the type and use of building that they serve. Wherever possible they should be provided close to the entrance and, especially on steeply sloping sites, every effort should be made to provide accessible parking at the same level as the entrance to the building. Each bay should display the International Access Symbol on the road surface. The symbol should be in yellow or white and a minimum height of 1400 mm.

For car parks serving buildings to which the public have access it is recommended that the following number of accessible car parking spaces are provided.

Workplaces   If the number of disabled employees who are motorists is known, provide one for each disabled employee who is a motorist plus at least one space or 2% of capacity for visiting disabled motorists (whichever is greater).

If the number of employees who are disabled motorists is not known provide at least one space plus 5% of total parking capacity (to include both employees and visitors), whichever is greater.

Shopping, recreation and leisure facilities   One space for each employee who is a disabled motorist plus 6% of total capacity for visiting disabled motorists.

Recommendations for railway car parks, churches and crematoria are given in BS 8300:2001

10.13 Emergency pull switch

Accessible toilets should be provided with an emergency assistance red coloured pull cord sited so that it can be operated from either the WC or the floor area nearby. The pull cord should be provided with two red bangles of 50 mm diameter, one positioned at between 800 mm to 1000 mm and the other at 100 mm above finished floor level.

Within the accessible toilet there must be both visual and audible feedback that the alarm has been activated once the cord has been pulled. This is also useful to advise the users if the alarm has been activated by accident, perhaps if it is confused with a pull cord provided for the light.

The alarm may be connected directly to a staffed position, for example a reception desk or security office, or may activate an alarm indicator outside the toilet itself. A reset button for the emergency assistance alarm must be provided which can be reached when seated either on the WC or in a wheelchair. A clear written procedure must be in place to state procedures that will be adopted when the alarm is activated. Staff must be fully conversant with the correct procedure to adopt and, if the alarm is sited adjacent to the accessible toilet, a notice clearly indicating what should be done if the alarm is activated should be placed on or near the toilet door. Simply opening the door from the outside without properly checking that the alarm has not been activated inadvertently could cause extreme embarrassment to the user and is not an acceptable management practice.

Figure B.2

Figure B.3

Figure B.4

Lydia House, administration centre, office building not accessed by the public

Vertical circulation

Staircases
There are two main staircases in the building. One gives access to the first floor adjacent to the meeting rooms and the other gives access to the area of the first floor adjacent to the goods lift.

The main staircase (Figure B.2) does not have nosings to the steps that contrast adequately with the rest of the step. The handrail provided, whilst of an acceptable diameter (40 mm), is fixed to the wall with brackets that are the same diameter as the handrail (Figure B.3), a design feature which could cause injury to users of the stairs and is not recommended good practice. The handrails do not extend beyond the top and bottom step and are not continuous around landings.

The staircase adjacent to the goods lift (see Figure B.4) has nosings that can be seen when descending the stairs but are less well contrasted when ascending the stairs. However, providing the nosings are kept clean, we do not envisage that this will present a problem on this staircase. The handrail does not extend beyond the top and bottom step and is not continuous around landings.

Lifts
There is no passenger lift in the building although there is a goods lift. We would recommend that investigations are carried out to see if this lift could be converted to passenger use.

Horizontal circulation

Corridors and circulation routes
The floor finishes at ground and first floor levels are highly reflective in places which, with glare from overhead lighting or windows and doors at the ends of corridors (Figure B.5), will be disorientating for some users, and especially those with poor or restricted vision.

At the time of the audit there was a considerable amount of 'clutter' and items stored in the circulation routes (Figure B.6) that

Figure B.5

Figure B.6

Figure B.7

represent tripping and collision hazards and should be removed as soon as possible. Monitoring and management procedures should be put in place to ensure that circulation routes are kept clear at all times.

Internal doors on circulation routes do not have glazing panels that extend down to 900 mm above finished floor level. This should be attended to at the next refurbishment stage or when a disabled person is employed at the Centre.

Some door widths at first floor level have a clear opening width of 700 mm rather than the minimum recommended of 750 mm (see Figure B.7). This will need to be addressed if a disabled person is employed at the Centre or if a lift is provided to make the first floor of the building accessible to wheelchair users.

Sanitary facilities

Accessible toilet facilities are provided in the male and female toilets at ground floor level and, perhaps somewhat bizarrely, at first floor level – although there is no passenger lift facility to the first floor.

In general, the design of the toilets does not meet even the minimum standards of Part M of the Building Regulations in terms of size, position and extent of the facilities provided within the toilets. The facilities that have been provided, such as the drop-down handle, flushes and locks (Figure B.8) are difficult, if not impossible, to use easily and would present many disabled people with considerable problems.

It is not possible to wash and dry hands whilst seated on the WC and there is no emergency alarm system provided (Figure B.9). Colour contrast is also poor.

The lack of a unisex accessible toilet facility in the building means it is not possible for a disabled person to be assisted by a carer of the opposite gender – something which is often required.

The position or presence of accessible toilet facilities is not signed within the building. The building cannot be considered as currently providing toilet facilities suitable for disabled people. In the standard toilet facilities, colour contrast is poor (Figure B.10) and the turn taps should be replaced with lever style.

Shower facilities are not accessible to disabled people (Figure B.11).

Figure B.9

Figure B.8

Figure B.10

Figure B.11

Extract of audit recommendations

An extract of the audit recommendations for Lydia House is shown in Table B.2.

Table B.2 Lydia House administration centre access audit recommendations

Para. no.	Problems identified	Remedial work required	Priority	Linked to maintenance schedule	Cost (£)
Vertical circulation – staircases					
3.1	Nosings to main staircase are not sufficiently contrasted	Improve contrast. See ICI Dulux Guidance	B	Yes	ii
3.1	Handrail to the stairs is not continuous around landings, does not extend beyond the top and bottom step and has inappropriate wall fixings	Provide an appropriate handrail to both sides of the stairs in accordance with BS 8300	B		iv
3.1	Nosings to the stair adjacent to the goods lift are not adequately contrasted when ascending the stairs	Keep nosings clean	A	Yes	0
		Replace at next refurbishment	B	Yes	i
3.1	Handrails to the stair adjacent to the goods lift do not extend around landings or project beyond the top and bottom step	Provide new handrail extending around landings and 300 mm horizontal beyond the first and last step	B		ii
Vertical circulation – lifts					
3.2	No passenger lift to first floor	Explore the possibility of providing a passenger lift	A	Working practice	0
		If structurally possible, provide passenger lift	B	Yes	v/vi
Horizontal circulation					
7.0	Reflective floor finishes reduce the effectiveness of contrast between walls and floors and present visual confusion	Change cleaning regime so the vinyl floors do not have such a highly polished finish	B	Yes and management practice	0
7.0	Glare from lighting in corridors	Upgrade lighting to reduce glare	C	Yes	iv/v

Table B.2 (continued)

Para. no.	Problems identified	Remedial work required	Priority	Linked to maintenance schedule	Cost (£)
7.0	Vision panels in doors on circulation routes are not appropriate	Replace doors with vision panels that extend from 500 mm to 1500 mm affl	B or as part of a workplace assessment	Yes	i per door
7.0	Some doors at first floor level have a clear width of 700 mm	Replace with doors with a clear opening width of 750 mm	C or as part of a workplace assessment	Yes	iii/iv
7.0	Considerable 'clutter' and obstructions in corridors at both ground and first floor level	Remove 'clutter' and obstacles, set in place management procedures & monitor	A	Management practice	0
Sanitary facilities					
8.0	There are no suitable accessible toilet facilities available in the building	Provide at least one unisex accessible toilet facility	Design as part of a workplace assessment		v
8.0	No appropriate accessible toilet facilities are available at first floor level	If a passenger lift is provided, provide a unisex accessible toilet at first floor level	C	Yes	v
8.0	The shower facility that is provided is discriminatory in that it is not accessible to disabled people	If a disabled person is employed at the Centre, provide an accessible shower facility	C or as part of a workplace assessment	Yes	iii/iv
8.0	Contrast to standard toilet facilities is poor	Improve colour contrast	B	Yes	ii/iii
8.0	Taps to basins in standard toilet facilities are twist/turn style	Replace with single-lever mixer taps with controlled water temperature	B	Yes	iii
8.0	Signage is poor	Improve signage to locate/identify toilets in accordance with the Sign Design Guide	A	Yes	i

Oliver House, office building

Table layout

Column 1   This identifies the location or specific item.

Column 2   This contains a brief description of or comment on the existing situation.

Column 3   This contains recommendation for improvement.

Column 4   This column is for the use of the designer to provide information on the action taken to improve access further to the recommendation. This can also be updated on an ongoing basis as maintenance, management and design improvements are implemented. Dates of implemented actions should also be added here.

Example

An example of the tabular format is shown in Table B.3.

Table B.3 5.0 Ground floor level

Column 1 Item	Area	Column 2 Current situation	Column 3 Recommendation	Column 4 Action
5.1	Ground floor	The ground floor is all on one level affording wheelchair access and easy travel for people with mobility and sight impairments		

Table B.3 5.0 (continued)

Column 1 Item	Area	Column 2 Current situation	Column 3 Recommendation	Column 4 Action
5.2	Reception	Indicative reception desks are shown on the plan. No detail available	Ensure low counter with high sections or vice versa, with leg space for close wheelchair approach to staff and customer sides. Ensure good lighting to maximise lip reading. Provide a minimum 1500 mm wheelchair turning circle to the staff side	
5.3	Cloaks	Cloakroom shown with service counter	See recommendations on counter design above	
5.4	WCs; Core G3	There are visitors WCs adjacent to the main entrance and reception area. These comprise male and female provision and a separate unisex accessible WC.	It is recommended that there is a WC designed for use by ambulant disabled people in male and female provision. See BS 8300 section 12.4 for detailed design guidance	
		There does not appear to be any nappy-changing facilities in this WC core.	It is recommended that there is provision made for nappy changing. It is recommended that this is in an area accessible to both male and female visitors. It is not recommended to locate the nappy-changing facility in the accessible WC, as when in use this will prevent disabled people from using the WC. Ensure good level of contrast in internal finishes in all WCs, to assist people with visual impairments to identify fixtures and fittings. Ensure taps are lever type. It is recommended that all WC layouts are reviewed in detail prior to construction	

Column 1 Item	Area	Column 2 Current situation	Column 3 Recommendation	Column 4 Action
5.4.1	Accessible WC: Core G3	There is one accessible WC for visitor use in core G3		
		The accessible WC measures 2 metres by 1.5 metres and is shown as having the standard Part M layout	See BS 8300 section 12.4 for design guidance on layout and fittings. Note that BS recommends a minimum dimension of 2.2 metres by 1.5 metres. Install alarm system in accessible WC linked to staff areas. See also notes for standard provision WCs	
5.5.1	Accessible WC: Core G2	There is one accessible unisex WC in core G2		

Appendix C
Information sources

Guidance

Access Audit Price Guide BCIS, 2002
Access for Disabled People Sport England Publications, 2001
Accessible Thresholds in New Housing: guidance for house builders and designers TSO, 1999
Building Sight: a handbook of building and interior design solutions to include the needs of visually impaired people RNIB, HMSO, 1995
Code for Lighting CIBSE, 2002
Designing for Accessibility CAE, 1999
The Disability Discrimination Act: Inclusion: a workbook for building owners, facilities managers and architects RIBA, 1999
Designing for Spectators with Disabilities Sport England Publications, 1992
Disability: Making Buildings Accessible Special Report Workplacelaw Network, 2002
Easy Access to Historic Properties English Heritage, 1999
European Concept for Accessibility CCPT, 1996
Guidance on the Use of Tactile Paving Surfaces DETR, 1998
Inclusive Buildings Blackwell Science, 2002
Inclusive Mobility: a guide to best practice on access to pedestrian and transport infrastructure DfT, 2002
Planning and Access for Disabled People: a good practice guide ODPM, 2003
Sign Design Guide Sign Design Society & JMU, 2000

Legislation, standards and codes

Copies of the following legislation are available from The Stationery Office. The legislation can be viewed at www.legislation.hmso.gov.uk

Disability Discrimination Act 1995 (DDA) The Stationery Office, 1995.

This is the principal statute. Note that this was varied by the Disability Rights Commission Act 1999 and will be substantially amended by Part II of the Special Educational Needs and Disability Act 2001 as it is brought into force (which is to be done piecemeal between September 2002 and September 2005).

The greater part of the DDA is now in force, and most of the remainder will be brought into force on 1 October 2004. For details of when individual sections of the DDA were brought into force, see the table printed at the end of the Disability Discrimination Act 1995 (Commencement No 9) Order 2001 (SI 2001 No 2030).

Disability Rights Commission Act 1999 (DRCA) The Stationery Office, 1999

This set up the Disability Rights Commission (DRC). The major significance of the DRCA for our purpose is the responsibility that it places on the DRC to draw up and keep updated the Codes of Practice on behalf of the Secretary of State. The DRCA additionally amended certain parts of the DDA.

Special Educational Needs and Disability Act 2001 (SENDA) The Stationery Office, 2001

Part II of this Act, which is to be brought into force section by section between September 2002 and September 2005, will extend the obligations not to discriminate to schools, higher and further education institutions and other providers of educational services. The act will insert new sections 28A to 28Z into the DDA and make a variety of other minor changes to the DDA.

Statutory Instruments

Statutory Instruments under the DDA can be viewed at www.legislation.gov.uk

The Codes of Practice relating to the DDA are available on the DRC website www.drc-gb.org

Code of Practice, Rights of Access, Goods, Facilities, Services and Premises The Stationery Office, 2002

Code of Practice, Elimination of Discrimination in the Field of Employment against Disabled Persons or Persons who have had a Disability The Stationery Office, 1996

Code of Practice, Duties of Trade Organisations to their Disabled Members and Applicants The Stationery Office, 1999

Approved Document M access and facilities for disabled people DETR, The Stationery Office, 1998

BS 8300:2001 Design of buildings and their approaches to meet the needs of disabled people – Code of practice BSI, 2001

BS 5588:Part 8:1988 Fire Precautions in the Design, Construction and Use of Buildings – Code of Practice for Means of Escape for Disabled People BSI, 1988

BS 9999-2 Code of Practice for Fire Safety in the Design, Construction and Use of Buildings draft, BSI

Human Rights Act 1998 The Stationery Office, 1998

The Human Rights Act 1998 came fully into force in October 2000. It incorporates into UK law rights and freedoms guaranteed by the European Convention on Human Rights. Some of these rights may have significant implications for disabled people such as the right to education, right to life, respect for private and family life and protection from inhuman and degrading treatment.

Equal Treatment Directive 1975 (amended 2002)

The Employers' Directive 2000 covers employment and vocational training. It prohibits discrimination on the grounds of sexual orientation, religion, disability and age. It is intended that regulations will come into force in October 2004 to amend the DDA to comply with the directive. It is proposed that these will include removing the exemption for small employers and bringing many of the occupations currently excluded within the provisions of the Act.

Useful organisations

The Access Association
Tel 01922 652010
www.accessassociation.co.uk

Ann Sawyer Access Design provides consultancy services, including access audits, appraisals and training.
Tel 020 8444 2311
e-mail a.sawyer@blueyonder.co.uk

Building Cost Information Service Ltd (BCIS)
Tel 020 7695 1500
www.bcis.co.uk

British Standards Institution (BSI)
Tel 020 8996 9000
www.bsi-global.com

Centre for Accessible Environments provides information and advice, training and consultancy services.
Tel and minicom 020 7357 8182
www.cae.org.uk

Center for Universal Design: a research, information and technical assistance centre based at North Carolina State University.
www.design.ncsu.edu/cud

Chartered Institution of Building Services Engineers (CIBSE)
Tel 020 8675 5211
www.cibse.org

Department for Transport, Mobility and Inclusion Unit
Tel 020 7944 3000
www.mobility-unit.dft.gov.uk

Disabled Persons Transport Advisory Committee (DPTAC) advises the government on access for disabled people to transport. Their website has a useful access directory.
Tel 020 7944 8011
www.dptac.gov.uk

Disability Rights Commission is an independent body established by Act of Parliament to eliminate discrimination against disabled people. It provides information and advice, supports disabled people, campaigns to strengthen the law and provides an independent conciliation service for disabled people and service providers. The DRC also writes and produces the Codes of Practice relating to the DDA. DRC helpline tel 08457 622633

www.drc-gb.org

The Disability Unit at the Department for Work and Pensions
www.disability.gov.uk

Employers' Forum on Disability
Tel 020 7403 3020
www.employers.forum.co.uk

English Heritage
Tel 0870 333 1181
www.english-heritage.org.uk

International Institute for Information Design (IIID): research and practice in optimising information, information systems and knowledge transfer.
www.iiid.net

Is there an accessible loo? (ITAAL) provides information on accessible WCs and publishes 'The English Directory of Accessible Loos'.
www.itaal.org.uk

JMU Access Partnership
Tel 020 7391 2002
www.jmuaccess.org.uk

National Register of Access Consultants: a UK-wide accreditation service and register of access auditors and access consultants.
Tel 020 7234 0434; mincom 020 7357 8182
www.nrac.org.uk

RADAR provides general information on the needs of disabled people.
Tel 020 7250 3222; minicom 020 7250 4119
www.radar.org.uk

Research Group for Inclusive Environments undertakes research and provides consultancy and training.
School of Construction Management and Engineering, The University of Reading, Whiteknights, PO Box 219, Reading RG6 6AW
Tel 01189 316734; minicom 01189 864253
www.reading.ac.uk/ie

Royal Institute of British Architects
Tel 020 7580 5530
www.architecture.co.uk

Royal Institute of Chartered Surveyors
Tel 020 7222 7000
www.rics.org

Royal National Institute for the Blind (RNIB) provides help, advice and support for people with visual impairments.
Tel 020 7388 1266
www.rnib.org.uk

Royal National Institute for Deaf People (RNID) provides help, advice and support for people with hearing impairments.
Tel 020 7296 8000; minicom 020 7296 8001
www.rnid.org.uk

Royal Town Planning Institute
Tel 020 7636 9107
www.rtpi.org.uk

Sign Design Society
Tel 01582 713556
www.signdesignsociety.co.uk

Index